PRESIDENTIAL

MUSINGS

FROM THE

MERIDIAN

Presidential Musings

Reflections on the Nature of Geography
by Past Presidents of the Association
of American Geographers

from the
Meridian

Edited by M. Duane Nellis,
Janice Monk, and Susan L. Cutter

FOREWORD BY DOUGLAS RICHARDSON

WEST VIRGINIA UNIVERSITY PRESS
MORGANTOWN 2004

West Virginia University Press, Morgantown 26506
© 2005 by West Virginia University Press

First edition published 2004 by West Virginia University Press
Printed in the United States of America

10 09 08 07 06 05   9 8 7 6 5 4 3 2

ISBN 0-937058-89-0 (alk. paper)

Library of Congress Cataloguing-in-Publication Data

West Virginia University Press.
Presidential Musings from the Meridian: Reflections on the Nature
of Geography / M. Duane Nellis, Janice Monk, Susan Cutter.
xviii, 260p.   23 cm. [corr. rpt.]

1. Geography—Periodicals. Association of American Geographers.
2. Association of American Geographers—Periodicals. 3. Association
of American Geographers—Officers. 4. Geography—Study and
teaching—United States. 5. Geography—Philosophy. 6. Geography—
United States.
I. Title. II. Nellis, M. Duane. III. Monk, Janice. IV. Cutter, Susan.

IN PROCESS

Library of Congress Control Number: 2004114161

Cover design by Than Saffel
The cover typeface is Adobe Caslon Professional, drawn by Carol Twombly
from specimen pages printed by William Caslon between 1734 and 1770.
Text is set in Minion, an original Adobe typeface designed in 1990
by Robert Slimbach.

Table of Contents

✳ ✳ ✳

CHAPTER 3: *Geographic Education*

CHAPTER 4: *Career Preparation and Alternative Professional Paths*

CHAPTER 5: *Creating and Maintaining Strong and Healthy Departments*

CHAPTER 6: *Diversity Issues*

CHAPTER 7: *The Professional Society*

CHAPTER 10: *Geography and Public Policy*

CHAPTER 11: *National Visibility and the Public Role of Geography*

Chapter 12: *Ethical Issues*

Chapter 13: *Humorous Musings*

Chapter 14: *Concluding Thoughts*

Chapter 15: *Biographical Sketches*

❀ ❀ ❀

Foreword

GEOGRAPHY'S FIRST DRAFT OF HISTORY

✻ ✻ ✻

This collection of presidential columns offers a unique and fascinating perspective on the recent evolution of American Geography. Drawing on writings of presidents of the Association of American Geographers from the period 1967-2004 in the *AAG Newsletter*, it provides a chronicle of the inner workings of history-in-the-making at the AAG and in the discipline of geography.

These presidential columns represent not only a valuable reflection of the field at the time they were written but also record the anticipation of new directions to come. Their function, and occasionally their intent, was to help shape the discipline, not simply to describe it. Together, they form a wide-ranging synopsis of issues critical to geography's development over the past forty years. The topics addressed mirror the richness and diversity of geography itself, a diversity that so intellectually advantages and distinguishes our discipline to this day.

Looking back, many themes addressed in these columns arise consistently over time, and even resurface as issues today. At the same time, this book also documents a trajectory of dramatic change and real progress in many broad areas. Geography's key support institutions (the AAG foremost among them) have been strengthened over the years, its relevance within the university and in society are now the envy of many, and increasingly strong linkages between researchers in academic geography and those in other disciplines and other sectors (public and private) are now flourishing. Far-reaching new research themes ranging from critical social theory to GIScience emerged during this period as dominant intellectual forces in geography to interact with, enhance, and complement its traditional strengths. A sea-change also occurred in the role of women in geography, and this is an important part of the story of this book and its times, told in first person by some of the authors of that transformation.

A review of these columns also reveals the extraordinary commitment by AAG presidents over the years to the advancement and well-being of geography. As Executive Director of the AAG, I have seen this commitment richly demonstrated on a daily basis while working closely with many of these past presidents, each of whom serves a one-year term. We are fortunate as a discipline to have had the leadership of concerned and committed presidents, who might simply have enjoyed a ceremonial presidential tenure, yet instead have confronted the tough issues of building and strengthening the discipline, year by year (and column by column).

An intriguing quality of this book is the opportunity to watch this history of our discipline and its institutions unfold from close up, as presidents address the challenges of the day in geography. Collectively, these columns represent a "first draft" of our history as a discipline.

Another feature of these presidential columns is often their intimacy. Usually written quickly, on a monthly deadline in a frequently busy schedule, they address the AAG membership as one would in a letter to a good friend, or on occasion, to an errant family member. They convey a concern not only for our discipline, but for our collective family, and reflect that caring and desire to be of some real help that often characterizes private correspondence.

For the scholar of the discipline, or for the new student of geography, this book and these presidential columns provide not only an "insider's glimpse" into the recent development of our discipline, but an intimate acquaintance with many of the people who have helped lead and shape that history of geography. It provides a first draft of geography's history, written in the moment of its occurrence, and at the same time shaping its future.

Douglas Richardson

Chapter 1

INTRODUCTION

❋ ❋ ❋

The year 2004 marks the Centennial Anniversary of the Association of American Geographers (AAG). From humble beginnings in 1904 on the University of Pennsylvania campus, the AAG has grown to become one of the largest professional geography organizations in the world with more than 8,400 individual members and exceeding 10,000 members overall (including institutional members). The Association, while primarily made up of geographers from within the United States, has over twenty per cent of its members from other countries.

Beyond the AAG's premier journals, the *Annals of the Association of American Geographers* and *The Professional Geographer,* and the Association's one annual and nine regional division meetings, a major element of the AAG's communications is the *AAG Newsletter,* now published eleven times per year. The *Newsletter* has become a key source of information on developments within the Association and more broadly, on what is happening in geography in general. Started in order to enhance regular channels of communication within the AAG during the presidency of Clyde Kohn in November 1967, it wasn't until 1983 that AAG presidents began to utilize this forum to provide regular reflections on key developments and happenings related to geography and the AAG.

The impetus for this collection resulted from a conversation at an AAG Executive Committee meeting in August 2001 at Meridian Place among immediate past president Susan Cutter, President Janice Monk, and Vice President Duane Nellis. Reflecting on our own experiences (and soon-to-be experiences) and recognizing the importance of these columns in reflecting on issues, trends, and opportunities in geography over the past twenty-plus years, we decided to capture key highlights of these columns through this book and make the presidential musings more accessible to a broader community within our discipline. Every past president was asked to recommend to the editors three to five of their favorite columns which we then organized around twelve themes. We decided on a thematic approach rather than a chronological one to highlight both the consistency and diversity of themes. The resulting collection weaves a narrative of interest that begins with early reflections, then goes on to recurring themes related to geographic education, career preparation, to characteristics of successful geography programs, and geography and public policy.

Chapter two begins with a selection of the presidential columns that were irregular items in the *AAG Newsletter* prior to 1983, but provided presidents serving during this period with a way to address the membership on "hot topics" of the day. Many of these "hot topics" have remained with us, surfacing occasionally as key messages of more recent presidents, although some were more relevant to the specific historic context within which they took place. Borchert, for example, in 1969, addressed the membership on the AAG's Executive Committee's decision to move the annual meeting from Chicago to Ann Arbor, in "light of statements and actions, on the part of Chicago officials, which many of us consider contrary to free inquiry and to the advancement of intellectual pursuits." Zelinsky made more frequent use of the *AGG Newsletter* during his presidency to reflect on such themes as the "non-recognition" of geographers in the national scene, or lack of familiarity by those outside geography as to what a geographer represents in the scholarly community. Parsons, in contrast, in 1974 focused on the slackened academic job market, the rich intellectual diversity of geography, the lack of quality manuscripts for the main geography journals, and the financial challenges faced by the American Geographical Society. Still further, Mikesell provided his thoughts on the issue of the day within the Association—whether to have an open convention policy and the associated growth in attendance at annual meetings (with acceptance of all papers versus the previous policy of only accepting refereed papers). The issue of how to deal with the dramatic increase in papers being presented at the annual national meeting has been debated by many AAG Councils since this monumental (and some would say forward-looking) decision.

In 1978, toward the end of his presidency, Marcus reflected on three issues that impact the AAG and geographers periodically to this day—the need for long range planning, the importance of a strong membership base, and intra-disciplinary behavior. In these early musings, Berry followed Marcus with thoughts on the importance of geography as a discipline and its role in planning. Hart extended the discussion on the importance of geography through reflection on academic versus applied geography and the role of the AAG in serving both constituencies.

During the late 1970s to early 1980s, AAG Presidents offered key thoughts on strengthening the discipline of geography as well as creating a stronger position for the AAG at a time when geography programs were under attack from those who did not appreciate the discipline's crucial contributions to global understanding and human-environmental inter-actions. During this period Helburn provided important thoughts on how to head off distress calls from university administrators regarding geography. Morrill, in 1981, revisited a similar theme related to threats to geography programs, simultaneously promoting stronger partnerships among the AAG, the National Council for Geographic Education (NCGE), and the K-12 community in order to define geography standards, a theme that would recur and culminate in 1994 with the publication of *National Standards in Geography*. John Adams ruminated in 1982 on whether the AAG's publications were effective given the dire need for more geography in the United States. Lewis challenged the membership during his presidency to move forward aggressively to seize opportunities in basic education for geography, to educate the media about the value of a geographic perspective, and to be more fully engaged on our campuses nationally.

These early presidential perspectives and musings touch on some key themes that recur again and again in more recent years. At the same time, they point to a time when geography was under attack, but through sound leadership and action, was able to weather these difficulties, strengthen the discipline's position, and move forward in articulating geography's value and common bonds as well as society's needs for a geographic perspective.

From these early beginnings, the remainder of the book utilizes the now regular perspectives of AAG presidents in the *AAG Newsletter* to touch on a variety of themes that have been important in the transformation of the AAG and geography to what it has become today as the Association starts its next century. Chapter three focuses on geographic education with particular attention to teaching and learning geography in K-12. These reflections span a period when geography was gaining national visibility through the National Geographic Society's (NGS) successful efforts to develop a series of Geography Alliance Networks in every US state. NGS also provided support in partnership with the NCGE, the American Geographical Society (AGS), and the AAG during this time to develop the National Geography Standards in 1994. It was not unexpected, then, that Abler offered an early discourse on the need for strengthening education programs in geography (in 1986), followed by Wilbanks providing perspectives on the evolution of the standards in 1992, with Birdsall writing on the importance of building linkages in teaching geography between college and K-12 environments. Gober reminded us of the need for also reaching out to community colleges in order to strengthen geographic education, and the imperative and opportunities related to it especially in reaching minority students and potential majors. The teaching of geography was not lost and provided themes in columns by Hanson, Golledge, and Murphy as well.

Chapter four celebrates the breadth of pathways in geography and offers perspectives on career preparation, as well as the alternative ways in which geographers engage professionally outside of academia. Abler advocates the recognition of the value of "practicing geographers" and expresses concern that academic geographers undervalue these colleagues. Kates expands on this theme by touching on the value of independent scholars and the environment within which such geographers can survive. Brown attempts to broaden even further our perspective on scholarship, focusing on the challenge to and from geography as it relates to change, continuity, and discourse, emphasizing the need to focus on the research question rather than the methodology. Gober challenges geography programs to alter their PhD curricula to better prepare doctoral candidates for diverse career pathways, while Graf elaborates on the need for rethinking our geography curriculum so that it includes the appropriate set of skills. Complimentary musings by Monk suggest the need for geography to create a climate that recognizes and rewards professional service, and the importance of recognizing careers in complex and diverse ways. Nellis reminds the membership that many successful geographers have pursued pathways outside the normal academic track and yet have contributed in significant ways to our discipline.

Concern with the characteristics of healthy geography departments has permeated presidential musings for a long time. Highlights of these are included in chapter five. Most of these presidential thoughts reflect on individual president's past or current administrative experiences. Palm, Birdsall, and Olson, for example, write that departments should work closely with university administrators aligning departmental goals with college and university goals, while Mather, Birdsall, and Gober speak to the need for departments to be effective at external communication and linkages to the campus community. Nellis and Golledge provide a synoptic view of what should be the characteristics of healthy departments— such as having an engaged faculty, promoting student and faculty success, emphasizing good teaching, having a regular visiting lecture series, and facilitating an active geography student organization.

Another recurring and important theme for presidential columns and the AAG relates to diversity issues, which are highlighted in chapter six. While a Diversity Task Force was finally created by the AAG Council in spring 2003, interest in diversity and related themes has been on AAG presidents' minds for some time. The value of diversity from an intellectual and human resource perspective is addressed from a quantitative by Palm, who extends this discussion to strategies for the recruitment and retention of women. Hanson addresses multiculturalism from a content perspective, while Olson describes the need for affirmative action but also the privacy of individuals in self-identification. Then Golledge provides important perspectives on the role of retired geographers as an important human resource at the same time that he brings attention to the importance of disability issues. Monk further articulates the diversity theme through her discussion of who's teaching whom and how.

Throughout the history of the presidential columns, the AAG's leadership has felt the need to remind our membership of the importance of our professional society, and the ways in which the AAG interfaces with other such associations. Such reflections are highlighted in chapter seven. Demko, for example, muses on the role and linkages of AAG with other professional bodies (e.g. Consortium of Social Science Associations or COSSA). Jordan's perspective is focused more inwardly on the AAG itself as a professional society and its "intellectual mission." Clearly the AAG is an important home for intellectual discourse on many fronts. Ward and Brown provide some perspective on the AAG, its sister organizations, and histories of other professional societies. The role of regional meetings (Olson) and the role of specialty groups (Ward) have become an even more significant focal point for AAG members. The place of plenary sessions (Ward), the possibilities of alternative meeting formats for active learning (Olson), and the link between the AAG and our annual meeting and international relations (Monk) are expressions of the changing and dynamic nature of our annual meetings. The chapter concludes with the question posed by Murphy: What if there were no AAG?

Chapter eight reflects presidential perspectives on the importance of the discipline in strengthening international studies and broadening society's view of the world (Abler and Cohen). The role of regional geography and the importance of the discipline in international education gain considerable attention as they have created debate in the professional geography community (Cohen and Wilbanks). September 11, 2001, brought additional focus on the importance of geography in understanding our world from a geographic perspective (Monk and Nellis). And international perspectives have often reminded our presidents of the importance of regional geography (Murphy). A recurring concern in this chapter is how American geographers relate to their colleagues in other parts of the world. Issues addressed include the extent and nature of US participation in the International Geographical Union (IGU), the diverse ways in which American geographers could benefit from collaboration with and provide support to colleagues abroad, especially those with fewer resources than we have, and the implications of increased participation in the Association's annual meetings by geographers from abroad, especially western European (Cohen, Wilbanks, and Monk).

Chapter nine touches on individual presidential perspectives on where we should be headed as a discipline. Such attention has grown as geographers reflect on new impacts on the discipline of Geographic Information Systems and changing attitudes of a new generation of students—the millennials—and what geography can do to position itself to take advantage of these changing attitudes and opportunities. Many interesting perspectives are included in this book, which draw on lessons from our understanding of the Columbian encounter (Wilbanks), to new questions for the 21st century (Demko), to future geographies (Nellis). Recurring themes include concerns for fragmentation (Gober), specialization and overspecialization (Gober and Graf), integration and making connections within geography and beyond (Hanson, Graf, and Golledge). The essential role of synthesis (Brown) and geography as an integrative science capable of understanding big science questions are discussed as well (Cutter and Golledge).

Chapter ten examines the need for and importance of addressing public issues and participating in public policy debates (Cohen, and Kates). AAG presidents have been fastidious in reminding our membership of the importance of addressing public issues and the overall importance of geography in an uncertain world (Cutter and Murphy). They have identified a range of approaches that geographers could take in entering public arenas, from approaching governmental bodies (Wilbanks) to serving as expert legal witnesses and consultants to governmental and nongovernmental organizations and private agencies (Graf) to writing "op-ed" pieces for newspapers (Murphy). They have also challenged us to think about how we set research agendas, and about balancing the perspectives of being a "fixer" or a "critic" (Kates).

Chapter eleven further extends this policy dialogue in focusing on the need to enhance the prestige of the discipline and also reach a popular audience. Abler suggests some leverage points for enlarging the geographical community. Jordan implores geographers to combat anti-intellectualism within the discipline and advocates placing more emphasis on geography's intellectual core. He fears a rush to applied work will detract from the discipline's prestige. Cohen proposes strategies to reach intellectually elite students while Cutter argues for the critical need to promote reintroduction geography into elite college and universities such as the Ivy League institutions, which educate so many of the nation's business, political, and legal leaders. Jordan and Demko both call for more engagement with books publishing as a means of attracting wider audiences with our message. Wilbanks takes up the concept of rediscovery in detail, while Brown reviews the book *Rediscovering Geography* issued under the auspices of the National Research Council. Finally, Golledge reminds us of the reasons why we shouldn't be ashamed to call ourselves geographers.

Chapter twelve addresses both internal and external ethical issues and our role as professionals (Kates and Birdsall). These columns range from a code of ethics for geographers to addressing hate crimes (Kates), disciplinary tolerance (Birdsall), fakery in the publication game (Graf), and the ethics related to the overextended use of global positioning systems (Cutter).

Chapter thirteen sheds some true humor on presidential perspectives by helping us to lighten up (Kates) on geography at the same time that there are serious messages related to how to kill a department within a university setting (Abler) and how to conduct an unsuccessful faculty search (Cutter). The final chapter provides some reflections and feedback from members as seen in excerpts from some of the final columns of presidents with Olson's final column reproduced in its entirety. The editors also provide their own reflections on the volume.

Now read on and enjoy these presidential musings from the meridian reflecting on thoughts, ideas, and perspectives from AAG leaders on the most recent quarter of century of the Association's history. We hope you find these musings as interesting as we did in our rediscovery as part of our selections for this book.

❀ ❀ ❀

Chapter 2

BEGINNINGS (1967-1983)

❋ ❋ ❋

This chapter includes remarks from AAG presidents that appeared in the *Newsletter*, but were not regular features. In the beginning, Presidents most often communicated about pressing Association business, but near the end of the time period, their columns appeared more frequently and focused on broader themes of interest to the membership.

❋ ❋ ❋

CLYDE F. KOHN
November 1967

THE ASSOCIATION OF AMERICAN GEOGRAPHERS performs a wide range of functions. Some of these are externally oriented, such as the work of the newly organized Committee on Geography and Business. During the 1960s, the Association has become engaged more and more in activities of this kind, activities that relate the profession more closely to groups having interests in common.

More numerous, however, are the functions which the Association performs that are mainly internally oriented. One of these functions, and in my estimation the most important one, is to develop effective channels of communication between members of the organization. The Association performs this function, in part, by holding annual meetings, and by maintaining a publications program. As an additional means of communication, this monthly *Newsletter* is being instituted.

To improve communication channels with the organization is an important objective of your officers and Council members. Clearly more channels are needed, some of which can be designed, as this *Newsletter,* to serve the entire membership. But additional channels need to be established to serve the needs of the several highly specialized groups within the organization. Suggestions for implementing this objective are welcome, and will be given careful consideration.

JOHN R. BORCHERT
November 1968

DURING THE PAST FEW WEEKS, two to three hundred members of our Association have taken the initiative to urge that we withdraw our 1969 meetings from Chicago and to express their reluctance or refusal to meet there. These colleagues comprise a wide cross-section of institutions and scholarly interest. They take their position in the light of statements and actions, on the part of Chicago officials, which many of us consider contrary to free inquiry and to the advancement of intellectual pursuits.

Before it acted on this request, the Executive Committee polled the full Council and discussed the issue with as many people as possible including geographers in the Chicago area.

The Council divided ten to ten on the issue of moving the meetings to an alternative location. Subsequent discussions brought out a wide range of serious underlying considerations. I believe that a poll of the entire membership at this time would show a similar two-way division and a similar underlying diversity of thought.

It is the judgment of the Executive Committee that the Association must meet in 1969. Further, it is the Committee's judgment that a decision to meet in Chicago at this time would result in serious diversion of energies better directed into social geographic research and teaching which will help to illuminate and dispel the fundamental problem which concerns us.

Therefore the Executive Committee has taken unanimous action to remove the 1969 meetings from Chicago and to accept an invitation from the Department of Geography at the University of Michigan to meet in Ann Arbor, August 10-14, 1969.

We appreciate the invitation from Michigan. Our colleagues there assure us that the facilities will be adequate for a large and effective meeting, and we hope to see you there. I want to acknowledge for the Council, also, its most sincere appreciation of the personal communications and the thoughtful opinions and arguments, which many of you have transmitted.

WILBUR ZELINSKY
October 1972

THE QUALITY OF THE RESPONSE to my first Open Letter two *Newsletters* ago was excellent—so much so that I should like to try again, this time with an emerging issue that may loom quite large in the next few years.

Most of us who have weathered the past generation would probably agree that two dominant trends in American geography during the post-World War II period have been quantitative growth and qualitative improvement. Our ranks have grown steadily and rapidly, as has the number and size of college departments; and even though there are many who are less than thrilled by the rate or direction of change, the intellectual advances of our field have been substantial, in some ways even remarkable. But if we have fretted productively about turning out more of a better breed, little thought has been given to the equally important question: What kinds of geographers do we really need? The expansionist period has been almost single-mindedly one of creating more of an upgraded traditional product: the college professor, partial or full, plus the occasional government worker, planner, or cartographer as a byproduct. Few will question the need for further improvements in our teaching, research, and writing, or the value of current efforts in those directions. But are we, as a profession, fully and honestly confronting our obligations to society by simply replicating ourselves? Or, to take a narrow, materialistic point of view, are we, who happen to be teachers, being fair to our students by preparing them for a saturated market?

After all those years of a leisurely trickle of doctorates in geography (an annual average of about 70 during the 1960's), there was that quantum leap to some 172 in 1970-71, and there is every reason to expect an output of 300 per year by 1980, not to mention the great droves of master's degrees. We can hardly hope the number of job openings in colleges, much less those in elementary and secondary schools, will keep pace with such growth. But there is much else to be tended to in this world by bright young people with sound geographic training—the apparently endless ills of our urban agglomerations, of a polluted, badly battered ecosystem, the developmental strategy and tactics of the underdeveloped lands and our own impoverished regions and groups, of spatial injustice in all its forms, to note only the most obvious items. Shall we begin to think of ways of preparing geographers for meaningful careers outside academe and the federal establishment? How about allocating much of our growing manpower to a wide range of corporations (including the communications media), voluntary associations, citizens groups, local governmental agencies, or other forms of "applied" activity? What are the possibilities for hybridized geographers, those with advanced training in journalism, law, planning, engineering, ecology, or business administration? There aren't any answers available yet, but there is the prospect of serious discussion, perhaps at the Atlanta meeting next April. If you agree that these are important issues, may I have your thoughts and suggestions?

WILBUR ZELINSKY
February 1973

AT THE RISK OF BEING CERTIFIED as unduly thin-skinned or paranoid, I'd like to ventilate a problem that refuses to go away: the non-recognition of professional geography by the American public. Let me state precisely what I mean by "non-recognition." The term does not refer to prestige or our ranking in the scientific pecking-order. Virtually every scientific discipline feels that is it being shamefully under-appreciated, so that it is a waste of time and energy to lick our psychic wounds. Genuine status can be won only through good works and solid achievement.

I am fretting instead about the phenomenon of invisibility or sheer ignorance, that familiar blank stare that greets the geographer at a social gathering or an assemblage of scholars from other fields when he or she admits to being a geographer, then the patient explanation of what it is we try to do during our working hours. ("No, you didn't learn it all in the fourth grade. No, we are not mapmakers. No, we aren't exactly geologists either. Let me explain...") The inherent nature of our field does not induce such obscurity, and the problem seems much less severe in most other countries where geography has reached a high level of sophistication. If we were specialists in anything as arcane as phycology, paleography, or Finno-Ugric linguistics, we could resign ourselves to our unobtrusive niches. But I trust it is not unreasonable to wonder about the absence of that quick identification accorded the historian, architect, physicist, psychologist, meteorologist, or members of other broad fields. Has it been a matter of being bypassed by the coffee table book, the journalist, cinema, or television? Is it possible that our own ivory tower attitude has brought us to this pass—and perhaps also a disinterest in the folk geography of Americans, the "geosophy" of the common herd? And, quite apart from such mundane considerations as dollars and clout, should we worry at all about the causes and cures for this strange malady? Should we attempt anything in an organized fashion? A few of our colleagues have already speculated about the reasons, and some have offered some remedies; but little has been accomplished to date. I should welcome your thoughtful suggestions.

WILBUR ZELINSKY
April 1973

SERVING A ONE-YEAR TERM (or sentence) as president of a professional society is very much like parenthood, or life itself. It is only toward the end that you get the hang of the job, and by then, it's too late—and you're not sure you would ever want to try it again. In this valedictory letter, I'd simply like to share a few random thoughts, then quietly slip away.

This series of open letters seems to have worked well. The response has been splendid in terms of both volume and quality. In fact, I am still several weeks behind in reacting to the many thoughtful comments received. But whatever else may have been accomplished, I hope no unrealistic expectations have been aroused. Quite apart from teaching me again how finite my time and talents happen to be, these two years as Vice-President, then President of the AAG, have prompted a few fleeting thoughts on the character and role of the learned society in general, and ours in particular, and on the power and frailties of our Association.

I am thoroughly persuaded that the past quarter-century has been an era of remarkable, exhilarating growth in American geography, veritably our coming of age intellectually and organizationally. (I write these words well aware of all our growing pains and adolescent gaucheries and all the raucous laughter such a statement may stimulate among some colleagues; but it is true nonetheless.) The AAG may have well been the critical agent in catalyzing this growth. In fact, it may have overachieved. It is a far cry from the day when a few hundred keepers of the faith did little else collectively, but meet in solemn conclave annually and publish a relatively slim journal quarterly. Since then, the AAG may have become more of an activist, a stronger shaper of policy and direction for the discipline than has been the case for the central organization of almost any other social, biological, or physical science in the country. Our string of successes, which I won't rattle off here, may be exciting hopes that even the wisest, most efficient leadership cannot fully satisfy.

In any case, perhaps the time is approaching when the explicit definition of an optimal relationship between the national (or regional) organization on the one hand and our thousands of individual members and multiplicity of special-interest groups on the other—or between our discipline, the wider world of science and humanities, and the larger society to which we belong and are ultimately responsible—may claim an urgent place on our agenda. I cannot suggest the precise mechanism for such a searching review of how power, responsibility, and caring should be apportioned, but I shall be most interested in seeing what answers my successors can discover.

JAMES J. PARSONS
October 1974
WHAT ARE THE RESPONSIBILITIES of the incoming president of a professional organization such as the AAG? What might he or she do on accession to this privileged position to promote, even in some small way, the betterment of the field to which we are all in one way or another committed? One obvious possibility is simply to deliver oneself of a few personal reflections relating to the present state of geography as a profession. Permit me then the following:

1. The slackening of the academic job market, as well as some fairly significant structural changes going on within our discipline, has led to a recent and extraordinarily abrupt increase in emphasis on practical and applied geography, with the appropriate methodology and techniques. It is a move that has by no means run its course. We applaud and support it as a development that should significantly enhance our visibility and relevance to students, administrators, and the general public. The move towards a more applied form of professionalism is crucial not only for the maintenance of employment outlets for our graduates but also in facilitating the application of geographical insight to matters of public policy and concern. Yet it can be heady wine, and perhaps it is not too early to remind ourselves of the consequences should the imbibing go too far. Professionalism and scientific technique should supplement rather than supplant the humanism of geog-

raphy, the appreciation and understanding of the distinctive qualities of places, of people, of spatial arrangements. Only a completely "open geography" will assure this and protect from erosion the distinctiveness of our subject.

2.   The rich diversity of geography, which is so much its strength, has made it peculiarly vulnerable to academic encroachment. We have recently seen an intensification of this in the educational world with the rise of curricula under such now–familiar integrating rubrics as urban studies, regional science, earth science, (terrestrial science), man-land relations, ecology, environmental studies, natural resources and conservation, area studies, and a host of others. In general we have perhaps gained as many students as we have lost by such developments, for they have clearly helped to spread awareness of many of the themes closest to the heart of geography. Yet it behooves us all to be on guard lest our discipline—with its time-honored concern for the integration of human and earth sciences—be gobbled up by aggressive builders of academic empires. One too seldom considered way of defending our frontiers is through more imaginative course offerings, perhaps especially better and more attractive titling of courses, and more attention to field study as an integral part of geographical education. At an intellectual as opposed to a vocational level there is much scope, for instance, for local and problem-oriented urban and area courses, for making geography relevant to the rising tide of world travel and tourism, to the problems of resource adequacy and environmental quality, to regional differences on levels of development, to landscape appreciation as a humane art, to the history of land use and the accelerated, destructive exploitation associated with the imperfections in our economic systems, the demographic crisis, and the revolution of rising expectations. Where are our once–popular courses on world commodity production and trade? Why not courses on the coastal zone, on the oceans, on comparative agricultural systems, on the geography of social areas, open space, energy, retirement in an aging society? Or on issues relating to public transportation, land use zoning, housing, pollution, environmental hazards, endangered species, and habitats? A good course on landscape and literature, or place names, or climatic change, or the world food problem, or the geographical roots of social and political conflicts too, ought to pack the halls. Enrollment figures do count, for they are the measure of our

contacts with the students we serve and hope to positively stimulate as geographers. We cannot stand still, nor can we allow ourselves to be dominated by techniques at the expense of understanding, or we will be buried. (It would be interesting to hear of off-beat or non-traditional undergraduate geography courses—including changes of title or format—that have proven especially attractive to students and effective as teaching devices.)

3. Our professional journals are not receiving the number of quality manuscripts that befit our membership of 7,000. Given the disparate and fragmented character of our field it may be understandable that many geographers choose to publish in more specialized outlets not specifically identified with geography—journals devoted to area studies, economic development, natural history, ecology, earth sciences, or the more recondite forms of quantitative analysis or social theory. Yet we have a higher obligation to geography than our own personal self interests. If the subject is to prosper as an academic discipline we should all be directing at least some of our best work of more general interest and application to geography's own professional journals, especially those published and circulated in North America. This seems hardly too much to ask even if it must be at some modest personal sacrifice—usually no more than the cost of some extra reprints for specialists in other disciplines.

4. One of the most responsible positions in geography, the editorship of the *Annals*, comes open at the end of 1975. The AAG Publications Committee (Chairman: Richard E. Lonsdale, Department of Geography, University of Nebraska, Lincoln, NE 68508) is already considering candidates to be proposed to the Council and would appreciate suggestions or volunteers from the membership. The appointment is for a three-year term.

5. This country's oldest geographical organization, the American Geographical Society, is in dire financial straits and only an early and massive injection of aid will permit it to maintain the comprehensive program developed over the last half century. The Society's research and publishing program, including the *GR, Current Geographical Publications, Soviet Geography,* and all the rest, are threatened. So is the magnificent AGS library, a national resource that neither the country nor the profession can afford to see disintegrate. For areally organized information, especially, it is unique. The geographers of this country

could well search their souls for leads regarding sources of support or suggestions for ways in which the AGS might move to provide itself with the financial footing that it requires to survive. Bob McNee, the Society's Director, right now probably has the most crucial role of any person in American geography. He needs all the help, the ideas, the moral and financial support that we can collectively provide him.

MARVIN W. MIKESELL
October 1975

FOR THOSE WHO HAVE ATTENDED recent annual meetings, it is obvious that there has been a substantial increase in their size. This development is not a consequence of an increase in the size of the Association, which has been relatively stable during the past three years, but rather is the result of an "open convention policy." After the 1973 meeting, the Council directed the next Program Committee to accept all papers submitted according to its guidelines. The result of that policy was an unusually large program in 1974. Volunteered papers plus contributions to special sessions added up to a total of more than 300 presentations, nearly double the number offered in the previous year. The meeting in 1975, conducted according to the same "open convention policy" (and with the additional innovation of "poster sessions"), had close to 400 presentations. The next annual meeting promises to be of comparable size.

Most members of the Association are probably aware of the consequences of this policy, but it is not likely that the reasoning behind it (and the vigorous debates that have sustained it) are well known. The policy might best be described as an experiment, for arguments for and against it continue to be offered in Council meetings. Proponents have felt that a program open to all comers, or at least to those who are able to meet guidelines, is a healthy situation for the Association. It has also been asserted that a place on the program of the annual meeting might be regarded as a legitimate claim of AAG members, particularly younger members, for whom the right to be heard might mean the right to be hired. An additional argument offered is that the quality of the last two annual meetings does not seem to have been inferior to that of earlier meetings for which papers were more carefully screened.

Opponents of the idea of an open convention have argued that an attempt should be made to conserve the most valuable possession of our members: their time. A program restricted to a smaller number of presentations would also have the advantage of being less troublesome to those responsible for program planning and local arrangements. Programs requiring ten or more concurrent sessions create numerous logistical and even intellectual problems and could restrict our meetings to hotels of enormous size. Moreover, a policy of "anything goes" can be condemned as an abdication of responsibility.

Some papers or all papers? The Council continues to debate this issue and to review arguments pro and con. And it continues to be troubled by the virtues of a meeting that reflects the "real world" of American geography as against one that offers evidence only of what a committee regards as presentable products of our world. Brilliance and ineptitude, excitement and boredom, standing room only and nearly empty rooms—these have been the obvious contrasts of our recent meetings. But what is the purpose of our meetings? To provide a forum for presentations of exemplary quality or to offer a setting for display of the efforts of a much larger and more representative group of authors?

The fact that the Council has endorsed the idea of an "open convention," does not mean that this issue has been resolved. Debate continues. And it should be reinforced by the thoughts of a wide range of AAG members. The question of what kind of meeting we should have—or indeed can have, given our size and complexity—concerns all members of the Association. As citizens are often urged to write to their congressmen, so also should members of the AAG write or talk to their councilors. The recent gatherings in Seattle and Milwaukee demonstrated that there are advantages in a big, sprawling, and untidy convention. A smaller, more tightly controlled meeting would also have advantages. The councilors who must review these alternatives need to be aware of your views.

MELVIN G. MARCUS
June-July 1978

THIS WILL BE MY LAST OPPORTUNITY to address you officially as President of the Association, and I would like to take this occasion to share some thoughts with you. Three issues merit attention: (1) long-range planning, (2) membership, and (3) intra-disciplinary behavior.

Long-Range Planning

A major task of the long-range planning process was completed at the New Orleans meetings of the Association. At that time, and on schedule, the Report of the Long-Range Planning Committee was received by Council, and the Committee and its various Task Forces were dismissed with the thanks of the Association. The contents of an earlier draft of that report had been available to the Executive Committee in early March, and the Final Report was made available to full Council in late March. On the basis of those documents, the Executive Committee and Council were prepared to respond to a number of the recommendations at New Orleans. The details of those actions are contained in the Minutes of the 7 April 1978, Executive Committee Meeting and the 8-9 April 1978, Council Meetings, which appear elsewhere in this *Newsletter.* Please read these Minutes carefully as a guide to the status of long-range planning.

To put the Minutes in perspective, however, a brief review of procedures may be helpful. After reviewing the LRPC Final Report, the Executive Committee divided both formal recommendations and less formal suggestions into three classes: (1) items that could be acted on directly by the Executive Committee; (2) items requiring full response by Council in New Orleans; and (3) items which require extended consideration by Council and/or the membership. Perusal of the minutes will reveal actions relating to each of these categories.

Because of its length (70 plus typewritten pages), the actual Report of the Long-Range Planning Committee has not been reproduced for distribution to Association members. It is, of course, an open document, and any member wishing to read the full report may obtain it at copying and postage costs from the Central Office. Alternatively, members living in the vicinity of Council members or Long-Range Planning Committee members might be able to view a copy locally. In any event, we are encouraging members to maintain a dialogue with their councilors on this subject during the coming year. We hope that this year's regional meetings also will continue to provide a forum for long-range planning efforts.

The membership will be further informed regarding long-range planning via a series of commentaries in *The Professional Geographer.* The May 1978 issue presents an overview by me. Later issues will focus on results of the membership survey and topics considered by the specific Task Forces.

Although we have accomplished a great deal, the long-range planning process will require our serious attention through much of the coming year. Many promising recommendations require intensive follow-up by Council, Association Committees, and members. There is, I think, a tremendous potential in the suggested development of Special Interest Groups. This would help to ameliorate the problem of satisfying all the members all of the time. We come from a broad spectrum of intellectual, pedagogical, and interdisciplinary positions. Thus, Special Interest Groups may give us the opportunity to exercise special interests and at the same time participate fully in Association affairs.

Membership

In an earlier *Newsletter*, I appealed for membership participation in the long-range planning process. Part of my argument stated "For better or for worse, the Association is a product of its membership and the 'representativeness' of the Committee's final recommendations will be a function of which and how many geographers contributed ideas and opinions." That comment is worth reiteration in a slightly different light. Sometimes we forget that the Association of American Geographers is a democratically structured society. We each own only one vote and occasionally we are going to sit on the losing side when issues are decided by membership vote. Now my analogy may be a bit thin, but people don't give up citizenship and all the rights and services that they enjoy when tax levies are voted. They may not be very happy, but they usually stick around. Similarly, they don't run off in a corner when a referendum is passed that displeases them, but they remain to address the next issue. The not-very-subtle point of this is to encourage full and continuing Association membership and the renewal of past members in cases where members have resigned because they disagreed with a vote.

Executive Director Warren Nystrom reminded us in New Orleans that there is always a drop in membership after resolutions are passed or defeated, with losses incurred in either case. That membership base is only slowly rebuilt, and this takes its toll both on the Association and on the member who loses a significant professional connection. The Association will vote on a mail ballot in early autumn, which concerns the Equal Rights Amendment (see Minutes of 8-9 April 1978 Council Meeting and 11 April 1978, Annual Business Meeting elsewhere in this *Newsletter*). If the past foretells the future, we can count on membership resignations regardless of the outcome of that democratically-administered vote. Let me encourage you to make 1978 and 1979 the years in which we baffle the probabilities by retaining, and even increasing, our membership, while continuing the important dialogues that concern us as geographers.

Behavior Toward Other Geographers

"Geographers have a death wish." "When all was said and done, the reviewers from other disciplines were favorably impressed by the proposals submitted by geographers, but the reviewers who were geographers were extremely negative and destroyed the chances of the proposals receiving support." "During the deliberation on various cases, I became aware of the cohesiveness among members of other disciplines . . . I found this in distinct contrast to my experience with geographers, where intra-disciplinary feuding seems to be in vogue."

These quotes are examples of some of the comments I have heard regarding an issue that seriously concerns many geographers; namely, the tendency to put down our professional colleagues and/or our discipline. Now I am not objecting to appropriate and vital interaction and debate; rather, I am concerned about the small, but vocal, number of geographers who act destructively out of what seems to be pure pique or personal vendettas. Obviously, this is a problem which occurs in every field of endeavor, but geography, because it is a small discipline numerically, is particularly vulnerable. Thus, where the airing of a large English department's dirty linen might cause only embarrassment to those concerned, it can actually threaten the stability and even the existence of a geography department that is not so strongly established.

The prospects for geography are quite good these days. Employment opportunities have increased, and geographers are playing increasingly significant roles in both academic and non-academic arenas. At the same time, money is tight and competition for both government and private funding is heated. Thus, the discipline can ill afford petty, destructive actions. With this in mind, I ask that we all continue to operate in as fair and objective a way as possible and make every effort to avoid nasty and unnecessary *contretemps*. The latter profits no one and injures geography. Instead, let us, whenever it is appropriate, speak positively of our discipline and of our profession. We might even, every now and then, think of doing a good deed for another geographer.

My concluding comments are personal ones. It has been both a privilege and a pleasure to serve the Association during this personally most rewarding year. The workload at times seemed heavy; but, as I am sure my predecessors would verify, the President's way smoothly greased through the ministrations of the Executive Director, Warren Nystrom, Educational Affairs Director, Sam Natoli, and the rest of the Central Office staff:

Elizabeth Beetschen, Pat McKenna, Jane Castner, and Peggy Calaluca. They are the ones who keep the Association ticking along, and I am most grateful to them for their counsel and help. Also, I would like to express my thanks to the members of the Council and the various Committees who have been so helpful during the past year. Finally, I want to thank the membership for their letters of interest and encouragement, their warmth and hospitality as I visit different regions, and their kindness in electing me President in the first place.

BRIAN J.L. BERRY
August-September 1978

AN INCOMING PRESIDENT always approaches a year in office with some trepidation, and I am no exception. Mel Marcus guided us through an impressive year of long-range planning and presided over our biggest-ever annual meeting, with over a third of the Association's membership in attendance at New Orleans. Mel, Warren Nystrom, and the Central Office staff deserve our thanks for a job well done.

Yet much remains to be accomplished if the AAG is to remain efficient and effective in a changing environment. A new Executive Director must be appointed to replace Warren, who will retire after the Philadelphia meetings and to whom we should all be indebted for the extraordinary effectiveness of the AAG in recent years. We need to act on the question of Special Interest Groups to provide continuity and focus to the work of significant sub-groups within the Association. We must urge our colleagues who have dropped their memberships to rejoin the AAG, and encourage our students to become members. We need to explore closer ties with our sister organizations, examining the possibility of some broader form of confederation of America's geographical societies if it appears to be mutually advantageous. And as many of us as possible should plan to come to Philadelphia in 1979 to make the 75th anniversary meetings of the Association a gala affair, a celebration of geography, in keeping with the special theme of the meetings: the past, present, and future of geography. Many special events are planned, and the banquet will be reinstated for the occasion.

The year should provide all of us with the opportunity to reaffirm our commitment to the essential sprit and purpose of the field, that common sense of identity that unites us as a profession despite the plurality of our specialized research interests. Geographers seem to have a perennial uncertainty about their future, yet for those of us who spend much of our time outside the discipline, narrowly conceived, it is clear that there is a future, and an important one. Both the need and the opportunity are evident. In the fields of urban and regional policy analysis and planning, for example, there remains a critical need for the insights and skills of geographers, reaffirming the Scott opinion in England, which stated that "planning is the art of which geography is the science." From my position in a planning school let me also say this to our undergraduate teachers: geography graduates do extremely well in professional degree programs, and subsequently in professional practice, provided that their undergraduate programs have attended to the basics. We must ensure that Bachelors in Geography are both literate and numerate, and that they come to graduate school well-grounded in the conceptual and analytic core of the field; in physical geography, the environmental interface, and in human geography; in concepts of location and land use; and in cartography and spatial analysis. A demanding set of undergraduate requirements, we find, provides a better foundation for the average undergraduate than does the lassitude of unguided electives. The basis for performance must accompany the wish to do good. But with that basis, a world of opportunity exists.

I look forward to an exciting and productive year. I hope that we can all join together to make it so.

John Fraser Hart
August-September 1979

Some of the more awkward questions that promise to bedevil the AAG over the next few years revolve around activism and applied geography. These questions are not amenable to quick and easy answers, and they may well be insoluble, but they merit the benefit of your best thinking.

Should the Association assume a more active and aggressive political stance in attempting to further the best interests of some or all of its members? Some members shout "Yes!," and some scream "No!" We are almost equally divided on the issue, if the recent vote on the ERA boycott is any indication, and our differences are matters of opinion and belief that may not be susceptible to rational disputation. We feel strongly, even violently, about the desirability of greater activism by the Association, pro and con, and there are few signs of defection from either side.

I wish I had an easy solution for this problem, but I do not. Let us remember, however, that our professional colleagues are intelligent people of good will, no matter how muddle-headed they may seem on a particular issue. I hope we can respect and defend their rights to their beliefs, even when we disagree with them most vigorously. I hope geographers can maintain our tradition of being able to disagree without being disagreeable, of conducting our debates in a spirit of good humor, and that we can refrain from casting aspersions on or impugning the motives of those who may disagree with us. Please, let's keep talking to each other, even when we disagree.

Applied geography is a horse of another color. It is high time for us to stop talking about it and do something, if only to admit that a scholarly society can do precious little. We must stop wringing our hands and start facing up to some tough questions: does the AAG need applied geographers more than they need the Association?

To begin with, applied geography (and here I include the entire gamut of nonacademic geography) covers an incredibly diverse range of activities and employers. Academic geography, in contrast, is remarkably simple, even monolithic; most academics teach the same kinds of subject matter to the same kinds of students in the same kinds of classrooms, at whatever level, in whatever place. But what do a government geographer in Washington, an industrial location consultant in Chicago, and a county planner in California have in common, other than that all three hold degrees in geography, and that all three, bless them, continue to pay their dues to the Association? Their interests are so different that they find little to read in our journals, and they have even less incentive to submit manuscripts to said journals. They are not rewarded by their employers for scholarly publication, they are unaccustomed to writing scholarly articles, and who wants to invest precious leisure time in an endeavor that may result only in a nasty rejection letter from an ill-tempered editor? Forget it!

The academic geographer needs the leaven of practical experience that the nonacademic geographer has in abundance, but what can the academic offer in return, once the last degree has been conferred? Perhaps the AAG should accept the fact that it is a scholarly society with a "natural" membership of around five thousand academic geographers, and attempts to broaden its membership may vitiate its existing scholarly strength without providing a solid new base. We would be complete and utter fools, of course, if we tried to close our ranks to nonacademic geographers; we need them, and we should welcome with open arms any who wish to affiliate with us, but we should humbly recognize the fact that thus far we have been able to offer them little but good fellowship. What more can academic geographers offer that would make membership in their Association more attractive to nonacademic geographers?

What should the Association do? What should it try to be? The answer is up to you. I am always distressed when I hear someone refer to the AAG as "their" Association because it is our Association, yours, and mine, and it is going to do and be what you and I want it to do and be. It needs the active and dedicated commitment of all its members, and I think you will discover, as I have, that it can be remarkably responsive to anyone who is willing to pitch in, roll up his or her sleeves, and get to work instead of sitting back and carping about "them."

Nicholas Helburn
April 1981

From time to time departments in distress call upon the AAG. Sometimes the officers or the Central Office can help convince the administration that geography is an important discipline and that the department in question should not be phased out of existence. It is important that we put out these "brush fires," but our efforts in that direction have only short run value. In the long run the status of the field, both generally in the US and locally in any particular institution, depends on meaningful research in the larger universities and high quality teaching in the smaller universities and colleges.

Two recent articles in *The Professional Geographer* have dealt with these issues. Marvin Mikesell in the August 1980 issue asks "The 16,000,000 Hour Question" and discusses the justification for undergraduates spending those hours in geography. Tom Wilbanks and Michael Libbee in the February 1979 issue point to the danger signals and suggest strategies for "Avoiding the Demise of Geography in the United States."

We have every confidence that geography has made and will continue to make important contributions. If these are well known and understood there is no problem. In an atmosphere of retrenchment, however, "the free market place of ideas" may become distorted by traditional vested interests. It is worth an extra effort to make sure that both the "liberal" and the "vocational" values of geography are well presented.

There are a whole series of procedures that faculty can use in making sure that their image with the rest of the faculty and with the administration is appropriate. Good teaching and good publications must be called to the dean's attention. Departments should be alert to declining enrollments and to bored or alienated students. The sources of the problem need to be uncovered and dealt with even though faculty traditions make this uncomfortable.

It is important to share positive developments in your department with the rest of the profession. However it is equally important for departments to recognize potential threats to their existence. It is effective for a senior geographer to make points with the Dean and Vice President for Academic Affairs before anyone has "taken a stand," that is, before a recommendation to deemphasize or to phase out has taken place. Faculty must be assured of confidential and diplomatic treatment of such "early warnings." When a department is threatened, members or friends should get in touch with the AAG Central Office as soon as possible.

The AAG Central Office operates a departmental consulting service and maintains a roster of individuals who have been successful in representing the discipline in academic circles so that they may be called upon to visit deans and other administrators, preferably before a crisis develops. Further, we should be thinking about establishing a set of procedures that departments can use to marshal written and personal support from geographers in other departments; alumni; members of the National Academy of Sciences; representatives of geographical societies; Deans, Presidents, or Chancellors; members of various departments or agencies in the Federal Government; state and local government employees; and representatives of specialized and cognate fields such as geomorphology, climatology, hydrology, coastal studies, natural hazards, land use, planning, urban studies, women's studies, education, cartography, remote sensing, and others.

Of course, procedures will vary from case to case. Usually they should be orchestrated by the faculty in the affected institution. It must be kept in mind that these are merely tactics. Fundamentally we need quality research and teaching to justify our place in academe.

Richard L. Morrill
August 1981

I will not pretend that the coming year will be easy. The Association and the geography profession face a period of intense competition for less resources in the schools, colleges and universities, fewer jobs in the marketplace, and in many places a question of very survival. But I intend, and hope that all of us will, to act aggressively to enhance the status, role, and utility of geography at all levels in all arenas, and I welcome the advice of all as to the myriad ways this may be done. The AAG Central Office and Council will work hard to defend geography programs where threatened, but the best defense is quality, and we may, as an Association, need to think about quality standards and program evaluation and assistance.

Geography is understandably judged from outside by the quality of its publications, and this will receive the priority attention of the Council. We are also judged by the quality of our annual meeting. I have always favored maximum participation but on the other hand, we have a responsibility to ensure reasonable standards of performance, and will explore ways to improve our meetings.

Another priority concern is membership, because higher membership is the easiest way to keep dues down and create channels to new job opportunities. We need creative ideas on how to reach and hold much larger numbers of geographers who work in various levels of government, in business, industry, and in other non-teaching jobs. What are ways of providing recognition or other outlets for their achievements?

Geography is more vulnerable because of the weakening of its base in the school system. We intend to work closely with NCGE on the urgent matter of defining standards of geographic knowledge which pupils at various levels ought to have, and in publicizing these.

Finally, we are serious about achieving wider participation by members in the affairs of the AAG. The committee vacancies will be advertised, and all members are encouraged to volunteer to work in areas of their choice.

JOHN S. ADAMS
March 1983

GEOGRAPHY MADE NEWS in 1982 when a series of stories highlighted strengths and weaknesses of academic geography in the US The stories warned of trouble ahead unless Americans improve their knowledge and understanding of their country and the rest of the world.

There is no need to recite the statistics that describe the geographical ignorance of well-schooled Americans. Instead, we should ask what the AAG might do about this sorry state of affairs. After all, it's partly our responsibility.

Our Association certainly cannot take on every worthy task that is proposed, and it should not try. But there are some activities that remain central to our mission as a scholarly society—the activities that would be last to go if we were forced to trim our agenda. These include the annual meeting and the AAG publications program.

Several years ago an AAG Long Range Planning Committee proposed the creation of research specialty groups as a means of improving scholarly communication in geography. The AAG Council adopted the idea, hoping that specialty groups would add new luster, vitality, and direction to the annual meeting. The results to date, it seems to me, confirm the Council's judgment, but it will take another five years or so to achieve a fair assessment of the specialty group idea.

At present, the annual meeting seems to be doing its job nicely. Parts of the annual meeting deserve regional and national media coverage, and the Council has appointed a task force to identify ways to improve the flow of information to the media.

AAG publication programs deserve continued scrutiny. They form our largest single effort and include the *Annals, The Professional Geographer,* the *Resource Publications,* and the *AAG Newsletter,* along with occasional special publications. Basically the present configuration of AAG publications appears to be a popular one, but I think we must ask periodically: Is the AAG publication program the most effective one we could deploy given the resources currently devoted to it?

The answer must come from a clear sense of the mission of the AAG and an accurate appraisal of how the publications serve that mission. The mission was well stated in the original AAG constitution:

> "...*cultivation of the scientific study of geography in all its branches...by assisting the publication of geographical essays, by developing better conditions for the study of geography in schools, colleges, and universities, and by cooperating with other societies in the development of an intelligent interest in geography among the people of North America.*"

The current constitution uses different words but points in the same directions:

"...to further professional investigations in geography and to encourage the application of geographic findings in education, government, and business...by stimulating research and scientific exploration...(and) publication of scholarly studies...to aid the advancement of its members and the field of geography."

I believe that now is the time to discuss whether our present publications program is the most effective program for these times. At the Council meeting last fall I appointed a task force chaired by Bonnie Loyd and Thomas Saarinen to answer this question. Their charge is to examine the present mix of Association publications and evaluate that mix compared to possible alternative mixes of publications, including the possibility of a popular geographical periodical—perhaps one focused on American geography, not only on the geography of North America, but also on the work that geographers do in government, business, and the schools. As the 1982 news accounts showed, American geography has a story to tell—and is a story to tell—but too often the stories are kept under wraps, or never told at all.

Maybe this is the time to consider some fresh, bold steps. Send your comments to Loyd, Saarinen, or me.

Peirce F. Lewis: *A Call to Action*
March 1983

Professional geography today is a far cry from what it was just a short time ago. It's not that the content of the field has changed so much. Rather, the context within which we operate has changed. The AAG must respond to those changes in a vigorous, aggressive way. Unless that happens, all of us as individual geographers are likely to find ourselves in serious trouble.

Two of the changes are lamentably obvious in the course of recent events. In the first place, the field has suffered several major institutional setbacks that have jarred us all. In the second place (and obviously related to the first), recent press coverage of American geography has been less than flattering—and that has jarred us too.

But I see a third change as I travel around the country—and I have been doing a good deal of traveling lately. The setbacks have shaken many of us loose from any feeling of complacency that we may have enjoyed.

Geography, as a professional discipline, may be smaller than it was five or ten year ago, but it's a lot leaner and tougher too. In the long run, I think, that is likely to be an unmixed blessing.

Our opportunities are simply enormous. The reason is plain enough to anyone who is paying attention to what is going on in the United States today. All sorts of things are in flux—that only a while ago were thought to be fixed in place: economic conditions, political attitudes, social institutions. Times like these are always unsettling and sometimes downright disagreeable. But—and it's an important but—such times of change are also times of opportunity. If a tougher and leaner geographic profession is alert to those opportunities, the next few years can see unprecedented gains for the profession that we all love so deeply.

Right now, I see three sorts of opportunities. In my view, it is crucial that geographers recognize and grasp them—acting as individuals—and acting in concert through the Association.

First, we are going to see a major change in the way Americans go about educating their children. In some parts of the country, it has already begun to happen. President Reagan had said repeatedly that we must "return to basics," and—whatever one thinks about the President's educational credentials—a lot of Americans agree with him. It seems inevitable that Congress and many state legislatures will rouse themselves to write new laws—perhaps to raise more money for education—certainly by making major changes in school curricula. It is crucial (that word again!) that the Association of American Geographers be on hand when (not if) the Congress first sits down to define precisely what Americans mean by "basics." If the AAG is there conveying its message forcefully and articulately at the time new laws are being written, the position of geography in America's schools will be immeasurably strengthened for decades to come and we can begin to rebuild the shattered hulk of geography in the nation's elementary and secondary schools.

We (and by "we" I mean the AAG) cannot afford to miss that opportunity. If we do, American geography will continue to dwell in the educational shadows for another generation, and yet another generation of American children will grow to adulthood as geographic illiterates.

Second, we are already seeing changes in the way that news is reported in media all over the country. It makes no difference whether or not we like *USA Today*, or whether we like anchored show-biz masquerading as the evening news on TV. That's the way most Americans learn about the world these days, like it or not. We are painfully aware that geographers haven't done a very good job of cracking the old forms of media, but there are important opportunities to demonstrate our best geographic wares to whole nations by way of these new media that are still in flux, and that possess a voracious appetite for material. Again, we simply cannot fail to grasp these new opportunities. The cost to geography is simply too high to contemplate.

Third, we can seize the opportunities that arise. As an academic, I can speak most intelligently about our places of day-to-day work. I have visited enough geography departments in colleges and universities around the country to know that geography is healthiest in those institutions where geographers have individually and collectively grasped a myriad of opportunities which, in sum, make their discipline intellectually indispensable. I've visited a few such places lately, and geography is in a lusty condition: at the US Military Academy at West Point, at the University of Colorado, at the University of North Dakota. And there are plenty of others that we need to know about—and whose successful tactics we all might learn to emulate. Between now and next June, I'll be using this column to describe some of the ways that geographers can work to make themselves indispensable. And I'll describe some of the things that I think the AAG ought to start doing—in order to ensure the future health of our beloved discipline.

My own opinion is this—and the opinion is shared by my predecessor John Adams, and by our Vice President, Risa Palm: that the AAG needs to take new and aggressive action to ensure that geography becomes indispensable to Americans at every level of public and private life. I emphasize the word "aggressive," because I (and we) see opportunities for AAG action that we have not customarily entertained. These new efforts almost certainly will require that the Association find new resources—and that means money. But the long-run payoff for aggressive action is too great to ignore—and long-run penalties for inaction are too grim to contemplate.

PEIRCE F. LEWIS: *Geographers and the American Press*
March 1984

PROFESSIONAL GEOGRAPHY IN THE UNITED STATES does not enjoy a very good press. It's not so much that we have a bad press: front page articles in *The Chronicle of Higher Education* about geography's "fuzzy image" are fairly rare—fortunately. The problem instead is that the American press largely ignores professional geography, even when it deals with clearly geographical matters. It's almost as if we didn't exist.

Most of us find that frustrating. Partly, of course, it's a matter of wounded egos. Nobody likes to be ignored when they are doing what they think are important and useful things. But it's more than that—much more. Inadequate press relations are crippling to the profession—and damaging to us as professional geographers. We will never receive active general public support until we gain sympathetic public understanding. And that won't happen unless our professional activities are reported accurately and sympathetically in the public press. For us, as professional geographers, therefore, reaching out to the press should be a matter of primary concern—not something we do in our spare time, when we can't think of anything better to do.

There are two general avenues we can follow when we set about to mend our ways with the press. One route is through collective action, initiated and pursued by the Central Office of the AAG. The other is by individual geographers, dealing with the press as persons, or in small individual groups—through their academic departments, their government agencies, or their private companies.

The AAG has already begun to move—and I trust we will move even faster and more effectively in the near future. When the AAG Council recently redefined the job of Executive Director, it specifically mandated a high priority for better press relations. One of the Executive Director's major jobs will be "to ensure that professional geography receives frequent and positive coverage in the media." Clearly, whoever is installed as Executive Director later this year will confront relations with the press as a major part of his or her job.

In the shorter run, the AAG is now taking action to ensure that we put our best foot forward to the press at the Washington meeting in April, 1984. As one example of what's happening, I propose to initiate a practice that the American Association for the Advancement of Science had been using for years at its conventions with great success. A press desk will be placed conspicuously at the main entrance of the convention hotel, to be manned by experienced geographers who will be able to handle reporters' questions with ease and accuracy. At the same time, I have asked a subcommittee of the Program Committee to select a handful of representative papers to be reproduced and made available to reporters in press-ready form. I have requested the subcommittee to select papers with a subject and style that will command public interest. Thus, when reporters arrive to cover the convention, we will give them geographic work that we think meritorious and interesting—they will not be left to their own devices to discover eccentric research that would be of interest mainly to Senator Proxmire.

Along this same line, more than a few geographers have suggested that the AAG needs to hire a professional public relations firm to manage our publicity—to "sell" geography as one might sell toothpaste. Personally, I don't think that's a very good idea. In the first place, good professional PR people put a very high price on their services. If the AAG were to commit that kind of money, we might have little left to spend on anything else. Second (and more important), there is little to guarantee that a professional PR firm could do the kind of job we want done—simply because they don't know enough about the range and richness of professional geography. If professional geography is to be reported accurately in the press, the reports must come from people who know what they're talking about—and that means us.

Which leads to the second matter—of what we as individual geographers can and should be doing to improve the image of geography in the public media. Over the last few years, I've been paying fairly close attention to such matters, and I've noted a few things that seem to work, and more than a few that don't. Let me list a few of them:

1. We should stop denigrating "popular geography," as some academics have gotten in the habit of doing. It's a simple fact that you can't talk to the public unless you talk in language the public understands. That means writing and speaking in plain English. It means drawing clear, simple maps. Described in such terms, popular geography is simply a form of public education—and none of us needs to apologize for that.
2. We should not try to tout geography to the public as if we were selling breakfast cereal or depilatories. The best PR for geography—or for any academic field for that matter—is to tell the public about the interesting and important things that we know—and are in the process of learning. Historians are good examples of how to do that. Barbara Tuchman does not run about town crying "Let's hear it for history!" Instead, she writes fascinating literate books about the German invasion of France in 1914, or about Joe Stillwell and the American experi-

ence in China before and after World War II. The public—which reads Tuchman in best-selling lots—does not need to be told that history is interesting. Tuchman's history is interesting, and it speaks for itself.

The same thing applies to us: Geographers too often forget that we have a wonderful story to tell, a story that the public has never heard before, and lusts to hear. (It's no accident that *National Geographic* sells 11 million copies a month.) The moral is simple: we don't need to give the public lectures about geography. We need to give them geography.

3. We cannot expect the press to discover what we are doing unless we help them. In short, we must take the initiative—go to the press and tell them what we are doing. They are not likely to find it by themselves.

In my experience, there are several strategies and tactics that seem to work in seeking out the press.

Get to know working journalists on a personal basis, whether they be reporters, editors, TV announcers, or camera operators. At the primitive level, it's fun: good journalists are attentive people, and therefore interesting people. And it's useful too, since it's easiest to deal with people whom you know on a first-name basis, and whose special interests and needs you understand and appreciate. If you're on good terms with a reporter who has learned to trust your judgment, it'll be you he turns to for geographic information—not somebody else.

Look for what reporters call "news hooks"—special occasions when stories have special relevance. For instance, a story about the geography of tourism in the West Indies would be newsworthy before the Christmas holidays, but not in midsummer. A report about research in electoral geography might be very welcome just before November elections—but not afterward. Research on Karst topography is a hot subject if the national news has just carried a story about houses collapsing into Winter Park sinkholes. And so on.

When dealing with journalists, be reasonable and friendly—and positive about your subject. Contrary to general reports, reporters and editors are people, like any other people. To be sure, some reporters are cynical, and dishonest, and habitually discourteous. But most are not. It does no good (and much harm) to approach a reporter with a chip on your shoulder. Keep in mind that most experienced reporters are very tired of people who talk down to them, or try to manipulate them. The rule is simple, I think: if we want to present geography in a positive light, then we ourselves must be positive and honest about our subject—and treat reporters as the honorable human beings that most of them are.

Above all, professional geographers should stay alert to what the public wants—and I'm talking about the general public that reads the popular press and watches commerical television, but I'm also talking about non-geographers who read the more elevated journalism of the scientific press. In both cases, the non-professional wants the same thing: an interesting story, told in vivid, literate English. There's no reason why we geographers can't tell such stories—and do it in places that will do us and the general public a great deal of good.

✳ ✳ ✳

Chapter 3

GEOGRAPHIC EDUCATION

❋ ❋ ❋

Geographic education in all of its manifestations—the practice of learning, the teaching of content, and the outreach to create an informed citizenry—has been a focus of many presidential musings. Many of these columns highlight undergraduate educational programs, while others talk about exciting opportunities to improve geographic literacy in kindergarten through high school and community college.

❋ ❋ ❋

RONALD ABLER: *Of Basketball Coaches and Geography*
March 1986

DURING MY 18 YEARS at Penn State, three freshmen matriculated with declared geography majors. Why only three? Geography is rarely offered in primary and secondary schools. When it is, it is too often taught by uninterested or poorly prepared instructors. Who can inspire collegiate interest in a subject that is previously unknown?

The contrast in geographic education between the US and Canada is painful and instructive. Geography is distinctly and well taught in Canadian schools. As a result, geography is the second most popular major selected by matriculating students in Canadian colleges and universities. Ditto in the United Kingdom.

Who's responsible? We are, more or less. More US attention was devoted to school curricula and teacher training by our early leaders. After

1925, less attention was paid to such matters; nowadays, it is often the basketball coach, rather than a trained geographer, who teaches pre-college geography.

The Geographic Education National Implementation Project (GENIP), in which AAG has joined with the National Council for Geographic Education, the National Geographic Society, and the American Geographical Society, is now working to restore vivid, comprehensive geography to the nation's schools. No one should underestimate the size of this job. Rebuilding a coherent structure of concepts, courses, requirements, and teacher certification programs in fifty states could take decades. Who's responsible? We are.

For too long, and in too many colleges and universities, those with a serious interest in geographic education have risked professional suicide. We can't all specialize in geographic education, but the profession and the AAG must support those engaged in such efforts. The long-term benefits for all of us are worth squeezing this imperative in among our existing musts, shoulds, and coulds.

The new brilliant students finishing secondary school each year are not exposed to good geography when they are deciding what to pursue in college. The large number of students corresponding to those electing geography in Canada and the United Kingdom select other majors in the United States.

What could we do if more top-notch students entered our undergraduate programs in their first year? What could we do if the number of matriculating students electing geography were second only to the number electing history? What could we do if students of that quality and quantity were practicing and professing when they completed their degrees?

We can find out, and will, if we are willing to make a long-term commitment to sound school geography.

Susan Hanson: *A Valedictory Forbidding Mourning*
June 1986

There is good news and bad news for geography these days.

The bad news is that more reverses are likely among academic departments. Colleges and universities across the country are under continual financial strain. Geography programs—like those in other disciplines—will continue to be threatened.

Although the discipline and the Association must strenuously resist cutbacks and closures, we should also recognize that it is usually too late to save a department once it appears on an endangered list. Working to keep departments off such lists will yield better results than trying to save threatened departments.

The good news is that geography is thriving in some places, and in realms where it has long languished. New departments are being established; some existing departments are expanding. Government agencies and private firms increasingly appreciate the contributions only geographers can make, and the number of practicing geographers is growing accordingly.

As a nation we are now forced to interact with regions and peoples outside our borders—and increasingly on their terms rather than on ours—general curiosity about the rest of the world grows keener. In that promising environment, the nation's geographical societies have joined together to strengthen geographical education in the nation's elementary and secondary schools.

The long-term program to upgrade geographical education launched by the Association, the National Council for Geographic Education, the National Geographic Society, and the American Geographical Society, will, I believe, turn out in retrospect to be one of the most important programs for strengthening the discipline to emerge in this century.

The discipline's early leaders took great pains to ensure that quality geography was taught in the nation's schools. Our leaders began to neglect school geography in the 1920s, and this has resulted in the wretched state of geographical instruction in America's schools today. The postwar expansion of higher education enabled geography to develop graduate departments without a base of school instruction and undergraduate majors. We are now, belatedly, building a foundation for that academic superstructure, and I am convinced that geography will be much stronger—in government, in business, and in academia—when that foundation is in place.

These are exciting times for geography. I am grateful to have had the opportunity to serve as president of the Association this year. We do face serious challenges in the short term, and we would be foolish to underestimate their gravity. But I believe we will overcome them, and I am convinced that the discipline's long-term prospects are excellent.

SUSAN HANSON: *A New Year's Resolution*
June 1986

A TRIO OF ISSUES highlighted in recent articles in *The Chronicle of Higher Education* prompts me to propose a new year's resolution for academic geographers.

One of these is the predicted shortfall in the supply of college and university teachers expected to hit by the mid-1990s as retirements exceed the number of new PhDs. (The number of job ads in the past few JIGs suggests that perhaps this gap between supply and demand is already affecting geography.) The second concern is the recognition of the importance of teaching. As a prod to provoke the professoriate to ponder pedagogy, and to university administration to reward good teaching in promotion, tenure, and merit decisions, a report from the Carnegie Foundation for the Advancement of Teaching has proposed that the definition of scholarship be broadened to include teaching. The third is the report that undergraduate interests are increasingly shifting away from business and engineering toward disciplines that focus on social and environmental concerns.

At the intersection of these three news items lies, I believe, an important message for academic geographers: good teaching is essential to the renewal and vitality of our discipline, and now is the time to capture students' imaginations and idealism by conveying the conviction that geographic insights can empower people to make a difference. With the projections of a seller's market in academe, we need now more than ever to attract the best and the brightest to geography. With students now clearly wanting to contribute to solving problems like urban poverty, environmental pollution, uneven development, land degradation, and resource depletion, we have the opportunity to demonstrate the strengths of the geographic tradition in tackling problems like these. The key is effective teaching. Bright students who want to make a difference will become alert to the possibilities geography offers them if their geography courses are intellectually challenging and stimulating.

It is time to see that the strength of our discipline lies not only in pursuing new knowledge about the world but also in infecting others with an eagerness to pursue new geographic knowledge on their own. It is time openly to acknowledge and cherish and learn from the good teachers among us. It is time to reflect closely upon, and to share with colleagues, what happens on those days when our teaching "clicks," and we leave the

that glow that comes from knowing students have caught
enthusiasm. It is time to think anew about how people
t how we might tune our teaching to accommodate the
1ing styles in our classrooms. It is time to take advantage
gies that facilitate learning. In next month's column I'll
ink the AAG might help improve college and university
while, let us resolve to renew our commitment to teaching,
professional lives that is worthy of our best thought and

Susan Hanson: *Enhancing Undergraduate Geography*
February 1991

LATE THIS PAST FALL the ad-hoc Committee on College Geography
met in Washington to discuss the advisability and feasibility of establishing a new Commission on College Geography (CCG II). The first Commission on College Geography did much during its twelve year lifespan
(1963-1975) to invigorate undergraduate geography. Convinced that the
AAG should once again be playing an active role in undergraduate education, the ad-hoc committee (Bill Brown, Evergreen State College; Bob
Churchill, Middlebury College; Phil Gersmehl, University of Minnesota;
Jan Monk, University of Arizona, and Susan Place; California State University at Chico) ended its deliberations by recommending that Council,
at its April meeting, appoint CCG II and charge it to evaluate and enhance
college geography in the US. The tasks of such a commission might include reviewing the status of undergraduate education in geography, identifying existing course and curricular materials worthy of broad dissemination, developing new materials, identifying strategies to enhance active
learning and critical thinking in geography, establishing mechanisms for
diffusing materials, and involving faculty in the development, dissemination, and implementation of these materials.

Why CCG II now? The ad-hoc committee thought that contemporary social, institutional, and pedagogical changes suggest the need to
look anew at undergraduate education in geography. *Societal change*—The
committee identified three societal changes that deserve particular attention in the content of undergraduate curriculum: (1) global environmental change, requiring course materials on environmental issues from local
to global scales; (2) global economic, social, and political restructuring,
requiring course materials on economic, social, and political processes

from local to global scales; and (3) the increasing diversity of the American population, requiring course materials on multiculturalism from local to global scales. *Institutional Change*—At the institutional level, recent initiatives in K-12 geographic education are likely to bring geographically literate students to colleges and universities in larger numbers than we have seen before, posing the need to rethink undergraduate curricula. Also posing a challenge to curriculum design is the increasing tension between demands for liberal education and vocational training. *Pedagogy*—Educators now agree that settings that promote active cooperative learning, such as data-based problem solving, are more effective than those that are conducive to passive individualistic learning, such as traditional lectures. The effectiveness of different teaching strategies also varies with the sex and cultural background of the student. We need to rethink our course designs and learning materials in this light. Many faculty who, like me, earned graduate degrees in an earlier era may feel ill-equipped to design or teach courses on global environmental or economic change. Many would welcome opportunities to learn about and use new course materials on these topics as well as to learn about new technologies of GIS, personal computing, and satellite imagery that encourage active learning.

To meet these challenges posed by these societal, institutional, and pedagogical changes, the ad-hoc committee has recommended that the AAG, through a new CCG II, seek external funding to support developing course materials that would include the dissemination of materials and plans for faculty development. The AAG is uniquely positioned to play a pivotal role in the enhancement of undergraduate education in the US. Please write me if you are interested in becoming involved in the Association's work on undergraduate education.

THOMAS WILBANKS
August 1992

THE MOST IMPORTANT DEVELOPMENT for geography in the United States in many years was the decision by the nation's governors and the President in 1989 to list our field as one of five core subjects for education in America's schools, as part of a major national commitment to educational reform. This decision puts geography on the same level as English, mathematics, science, and history in looking toward the future of education in the United States.

Many of you are aware of this development from Osa Brand's piece in the April 1992 *AAG Newsletter* or mine in a recent *AGS Newsletter*, or from hearing about it at the AAG meeting in San Diego, or simply from being plugged into the educational reform process in your state or locality. Just in case you haven't gotten the word yet, however, let me try to summarize what's happening and what it may mean for us in professional geography.

Since our National Education Goals were defined at the "Williamsburg Summit," geography's place as one of five core subjects has been incorporated into the America 2000 strategy developed in 1991 by President Bush and Secretary of Education Lamar Alexander, who was one of those governors in 1989. This plan calls for voluntary examinations in all five subjects, including geography, for all students in the United States in the fourth, eighth, and twelfth grades. It also calls for establishing world-class standards in these five subjects and for annual reports to the nation on progress toward reaching these standards.

Although this is a Bush administration proposal, the geography part of America 2000 is not basically a partisan issue. Governor Clinton co-chaired the governor's committee that developed the "National Goals Process," and its strength derives from demands from the states and the nation's private sector for a better-educated citizenry and labor force, not from demands by the federal government as such. Ross Perot's position is not as well known at this point, of course; but he has been a leader in educational reform at the state level, and he shares the concerns that have gotten the nation's business community behind the national goals.

This new status for geography in the United States is not yet assured. Others, from the social sciences (collectively and individually) to the humanities, have questioned "why geography?" if a fifth core subject is to be added to America's educational agenda. And most of our states and localities are having trouble finding resources to address four core subjects, much less a fifth. We need geographers across the country to support the proposal through their professional examples and their political advocacy.

But the process of institutionalizing geography as a central part of a national educational reform has already begun to move. For example, student competence in geography in the three grades is scheduled to be tested nationwide in 1994 at the state level as part of a National Assessment of Educational Progress (NAEP). The framework for the geography assessment, which will be the first widely-circulated interpretation of what geography has to offer to educational reform in this country, was approved by the National Assessment Governing Board on May 9, 1992; and I think this framework is a very creative step toward linking geography's traditions with society's needs. The NAEP assessment is not the testing of all students called for by the President and governors; but it is the first step in that direction, and its content will influence further discussions of how geography fits into the bigger picture. By the time you read this column, a companion effort will have started to define "world-class standards" for student competence in geography at the three grade levels, with the results to be delivered to the federal government by mid-1993. The AAG has been a key part of these efforts, working with other national geography organizations.

These kinds of activities raise an enormous range of questions, from the philosophical to the operational. If geography does, however, become one of the five core subjects in America's schools, K-12 (and considerable progress in that direction is very likely), I think you can visualize what is likely to happen. As the educational testing and standards efforts converge and the results of geography testing start to be reported at the substate level (e.g., for individual school districts), the relatively low scores in most of our venues are going to stimulate a demand for more and better geography teachers and materials. This demand will trickle upward as a demand for more geography courses in colleges, for teachers and for faculty to teach them, and for graduate departments to produce these faculty members.

This prospect is very good news for geography, especially in a time when so many of our universities are considering ways to cut departments and programs. There is no more potent argument for maintaining (or even increasing) support to geography, even in a time of scarce resources for higher education, than a National Education Goals Process which says that the nation at large wants geography to be recognized as a core subject for teaching and learning in this country and that, as a result, demands for geography at every level of our educational system are almost certainly going to be increasing.

But we need to be prepared to respond to this opportunity with sensitivity and humility as a professional discipline, because the emergence of grassroots support for geography is not because US business leaders and policy makers are impressed with the *Annals of the AAG* or the dissertations of our PhD students. We face a situation where what society is looking for from geography is not necessarily what many of us do. We can choose to react by responding to the needs around us, which will change us as a discipline in many ways, or by pulling back within ourselves to preserve disciplinary orthodoxy, which will mean that geography will fade quickly as a prospective core subject because that is not what people are looking for.

Is this a Faustian bargain? Personally, I think not. To me, it offers a vision of geography in the worlds of scholarship and life that is at least as exciting intellectually as it is operationally. Some of you may disagree; but you need to know that the choice is at hand, and you need to know that the resolution will not be entirely under our own control as professional geographers.

In my next column, I will offer some thoughts about the challenges we face in linking this emerging opportunity with the world of the AAG— and why, in fact, we have rather suddenly and unexpectedly arrived at this point.

Steve Birdsall
August 1994

One of my small pleasures as dean is to interact constructively with faculty across the full range of disciplines and compare their ideas and efforts with those in geography.

Last week I was talking with two of our history faculty about a project on which they and another colleague have been working. Several years ago, they organized a national conference on issues facing public school history teachers, they edited a book that arose from the conference papers, and they prepared and taught a series of statewide workshops intended to improve the content and methods of teaching pre-collegiate history in the United States.

All of this was reminiscent of our discipline's state Geographic Alliances, now active for more than a decade and involving hundreds of college and university faculty and many thousands of K-12 teachers nationwide.

Two things my history colleagues said caught my attention. First, they observed that as a side effect of their work other faculty in their department are now interested in pedagogical issues in history. I noticed this because faculty in departments nationally known for their scholarly excellence are rarely also active in public school education, but this condition seemed to be occurring here. There is at least one parallel in our own discipline with faculty in a highly ranked department also very active in geographic education.

Then one of my history colleagues asked if I had any sense of whether there has been any change nationally in the value that faculty, especially those at large research universities, assign to efforts to improve K-12 education. This is a critical question because its answer underlies both individual faculty decisions about how they use their scarce time and collective institutional decisions about rewards, such as salary and job security. Certainly this is a question we must address in geography, as well.

While I have little more to go on than intuition, deans are supposed to have answers, so I told him that yes, I thought recognition and credit for such activities were greater than they had been. It may still be too early to say whether the shift in interest within PhD–granting departments is a wavelet that will spend itself to no lasting effect on the inert beach of our institutions' missions, or whether we are experiencing a true sea change. There have been wavelets in the past, but I think we may be experiencing the beginning of a more lasting recognition at doctoral institutions of the value of faculty efforts to improve the content of pre-collegiate education. I recognize that my own view, past and present, may be limited, but the momentum appears to have shifted on this issue.

There are at least three reasons why a true sea change may be underway. First, there has been much interest by the public in K-12 education during the past decade and persistent criticism of public higher education institutions during these difficult budget years. Faculty and university administrators are making the connection between what we do and what the public expects from us. Geographers in higher education also have a special stake in the products of K-12 education; the appalling geographic illiteracy of most college students affects the demand for our courses and the interest in our field as a career. As a result, more geographers than ever are devoting good and serious effort to significant issues in geographic education.

Second, I wrote last month in this column about teaching and research and how these activities are two faces of a single coin. They should not be thought in opposition to each other. It may be that the growing interest in how best to teach to all ages the important ideas in our discipline and the new geographic methods is related to the gradual recognition that teaching and research are part of the same undertaking.

Third, the 1970s and 1980s saw a kind of demographic mixing among the different kinds of institutions. During many of those years, there were far fewer new positions available in doctoral departments expecting high research productivity from their faculty than the number of well-prepared and research-interested new PhDs being produced. Many of these bright, well-prepared new PhDs found positions in non-doctoral departments. As a consequence, there is less difference now than there used to be between the scholarly interests of faculty at different types of institutions than the missions of those institutions suggest. And some of these faculty have added their energies and intellects to the efforts of others already working on improving K-12 geographic education.

Our challenge now is to equilibrate the unquestioned value of scholarly work that leads to new theory, new answers, and new approaches with the dedicated application of the latest insights and methods to improve learning. We need to bring together the latest thinking about a given issue and the exciting transmission of that thinking at all levels of education.

As geographers, we must develop the criteria to identify and then give full credit to colleagues who are successfully improving the quality (and numbers) of the students we will teach, the students we will attract as majors, and the graduates who will take our places in the profession in the next few decades. We are not there yet, but we need to be. We are all partners in the enterprise of elevating geographic learning and geographic understanding. We ought to value excellence in constructive labors toward our shared goal, regardless of the institutional location or intended destination of the effort.

PATRICIA GOBER: *The Community College Juggernaut*
December 1997

MOST MEMBERS OF THE AAG, particularly those of my generation and older, were introduced to geography as residential students in four-year colleges and universities. This model of higher education, however, is declining in importance with the growth of community colleges. Many future college students will be exposed to geography in very different learning environments than the ones you and I experienced as undergraduate students. Consider the following:

1. In the fall of 1995, community college students made up 39 percent of the nation's enrollment in higher education; they represented 48 percent of all enrollment at public institutions. In some states like California, Arizona, Washington, Wyoming, and Florida, they are a clear majority of all college and university students.
2. There are more than 5.5 million students enrolled in US two-year colleges, 1.3 million of them in California alone.
3. Some 60 percent of the students entering the California State University System originate in the community college system. At my own institution, Arizona State University (ASU), the figure is 50 percent.
4. Community college enrollment has grown 42 percent since 1976; four-year college enrollment has grown only 19 percent.

Community colleges have grown rapidly because they offer an inexpensive, accessible, flexible, and generally high-quality product. An hour of credit from the Maricopa County Community College District, the feeder network for ASU, costs $34 per semester, one-third the cost of an hour of in-state credit at ASU. The ten local community college campuses, spread throughout our huge, sprawling metropolitan area, are more geographically accessible than our main campus. They offer far more early-morning, evening, and weekend classes than we do. Moreover, they teach courses like introductory physical geography to classes of 35 students with integrated laboratory sections. We teach the same course to 240 students subdivided into laboratory sections taught largely by incoming graduate students. It is little wonder then that 75 percent of our geography majors are community college transfer students. Community colleges provide the small class sizes and the hands-on environment at the lower division level that inspire new geographers.

Community colleges are an especially fertile ground for recruiting minority students to our programs. In 1995 there were more African–American students in public two-year than in public four-year institutions (588,200 versus 572,500). Among Hispanics, this disparity is even more pronounced. There were 590,300 Hispanic students in two-year institutions compared to only 346,800 in four-year institutions. Any meaningful strategy for achieving greater diversity in our profession demands that we pay serious attention to two-year colleges, the setting where many minority students are first exposed to higher education.

Most geographers at traditional four-year universities know surprisingly little about community colleges, given their staggering potential to generate new majors and raise the visibility of geography more generally. Few community college instructors belong to the AAG or attend AAG meetings, and most are too burdened by heavy teaching responsibilities to engage in the types of professional networks that bring us in contact with local geographers. Few of us can relate to the special challenges of functioning as a lone geographer in an exclusively teaching environment. As one community college teacher told me recently; "I'm the only full-time geographer on my campus. If anything needs doing, be it scheduling, developing new courses, writing grant proposals, or whatever, I'm the one. Sure gets lonely out there." Her interaction with colleagues at nearby four-year universities is one-way—from her to the universities. I suspect that this pattern is not unique.

Theoretically, GIS is an avenue for expanding geography's visibility in this large and dynamic academic market. And yet, Mike Phoenix, ESRI's Manager of University Services, worries that geography is missing critical opportunities to integrate GIS into the community college curriculum. After two years of responding to hundreds of community college requests for GIS support, he analyzed the sources of his contacts and was dismayed to find that only 10 percent were from geographers. The others came from a wide range of disciplines with CAD instructors in technology programs being the largest group. He was informed by one dean that geographers at her school were not interested, and she was going to provide another department with a computer lab and a new position to get going with GIS. I shudder when I think of the lost opportunities for geography.

What can we do to help? First and most importantly, we can stop ignoring our community college colleagues and acknowledge their pivotal role in delivering post-secondary geographic instruction. We are, in fact, partners in a system in which two-year colleges provide introductory course work and serve as seedbeds for majors, and in which four-year universities hone the skills and broaden the outlook of geography majors. Second, we can facilitate workable articulation agreements with the community colleges. The four-year universities in one large state initially were quite resistant to the idea of community colleges offering GIS training, arguing that it should be upper-division material. These barriers eventually came down as four-year institutions recognized the potential for community college GIS to feed geography majors to the four-year universities.

We can also involve community college instructors in planning activities for Geography Awareness Week and invite them to speak to and interact with our department faculty. In addition, we might work with them to develop new courses in GIS and other subfields, and encourage our doctoral students to teach courses at the local community college. My September column argued that PhD students would benefit from exposure to a wider range of professional experiences. Teaching a course at the local community college or working with a community college instructor to develop a new course in GIS, statistics, or qualitative methods exposes students to a different teaching environment, and it provides community college instructors with much-needed contact with other geographers.

The regional divisions of the AAG are a particularly appropriate vehicle for integrating community college instructors more fully into the geographic profession.

Their regional focus provides a less costly and less intimidating venue to present research, keep up to date on new ideas in geography, and meet local colleagues.

Jim Fonseca of George Mason University has done the profession an extraordinary service by chairing the AAG's Committee on Community Colleges. The Committee has organized community college sessions at the AAG meetings for the past several years; developed the Community College Listserv, which has over 60 members; and organized the Community College Network, the AAG's first Affinity Group. The Committee is in the midst of conducting a nationwide community college survey, under the direction of Joan Clemons, that will give us information about courses offered, teaching conditions, articulation agreements, and faculty needs.

The answer does not lie at the national level alone, however, and we must all help. Given the relatively small number of community college instructors who belong to the AAG and attend our meetings, we should reach out locally to community college instructors on issues of articulation, new course development, GIS, and other areas of mutual interest. As a discipline we can ill afford to ignore these rapidly-growing educational settings and the people who teach in them.

REGINALD G. GOLLEDGE: *Community Outreach*
October 1999

THE GEOGRAPHIC ILLITERACY of the US population generally is an established fact. It is also recognized that this illiteracy can be traced to the minor position that geography has played in the K-12 curriculum over the last 50 years or so. But it is acknowledged that the combined efforts of the four major geography associations are making headway in reducing this problem, particularly in terms of the contributions made in the Geography Standards and the Geography for Life K-12 curriculum. Other sterling efforts include the Geography Bee, Geography Awareness Week, Geography Bowls, and other local, regional, and national competitions. While some of these efforts require specialized organizational structures to run them, one of them—National Geography Awareness Week (NGAW)—is an event in which professional geographers could participate. I consider this a discipline-wide opportunity to which we should all respond.

Explaining your current research interests to a K-12 class may seem difficult.

But Geography Awareness Week gives us the opportunity to do just that. This type of community outreach is part of what is needed to change the disciplinary image to a more positive one that could be of benefit to our society and nation. Let me tell you how one department—that at the University of California, Santa Barbara—has used this opportunity.

The first year that NGAW was announced, the UCSB Geography Department called for volunteers from its faculty, staff, and students to visit local K-12 schools to talk about modern geography—particularly those facets that were being practiced at UCSB. There were about eight volunteers, some of whom eventually went to two schools. Local school district superintendents were informed of the dates of NGAW and asked to support our efforts by encouraging school principals to invite department representatives (faculty, staff and students) into their classrooms for a thirty to fifty minute presentation and discussion. The topics offered by the volunteers included: The Earth from Space; Mapping your Neighborhood; Western China; Snow Flakes and Ice Fields; Habitat of the California Condor; Maps and Exploration; Our Enemy the Earthquake; Geography of Costa Rica; Ethiopia; and other topics.

Each year thereafter, a call for volunteers to participate in NGAW activities was made. There was no shortage of excuses for not participating. Some, of course, are always legitimate—a person is out of town, at a conference, on fieldwork, has an important research deadline to meet, and so on. Many excuses, however, were but thinly veiled attempts to hide the reluctance and perceived discomfort associated with having to make themselves very clear to children or young adults. Some thought that the work they were doing was too complex to be translated for K-12 consumption. Our reaction to such disclaimers was immediate—"If you can't explain what you are doing in clear and simple terms, then you probably do not know what you are doing." This was a challenge that was hard to ignore.

In the following years, the number of volunteers and the range of topics for presentation steadily increased until in the early 1990s we had over twenty volunteers who went to more than forty different schools, and over thirty people visited over eighty classrooms (many of combined classes) by 1997. Those who participated found a certain amount of enjoyment and satisfaction in making their presentations. Even young children sometimes come up with the most amazing questions when faced with interesting material.

As the level of participation grew, so too did the variety of topics. Volunteers offered to talk about: Wave Action and How it Influences Local Surfing Conditions; Why the Ocean is Like a Layer Cake; Remote Sensing and the Environment; Mapping the Globe—How Global Information About Atmospheric, Land, and Ocean Conditions Can Be Compiled and Interpreted; Computer Cartography—A Computer Tour of Washington D.C.; Cognitive Maps of your Neighborhood; Hands–on Crayon Cartography; El Nino; The Nation's Parks and Forests; Cities of Ancient Mexico; Modeling the Spread of Brushfires; and many more technical, cultural, and physical topics. Visiting faculty and students talked about their homelands in Swaziland, Mexico, Brazil, France, Sweden, China, Australia, Japan, and other countries.

As more schools developed computer laboratories, GIS software was demonstrated and discussed; tactual mapping and comprehension of auditory landscapes were examined using innovative technology; computer simulations of brushfires and landslides (both common to the local environment) were displayed and discussed from a causal perspective. Migration patterns and school closures were popular presentations as did the nature and causes of drought and ways to combat it. Problems of pollution and hazardous waste disposal found their way into classroom environments.

After ten years, volunteers from the department were giving over eighty presentations to more than 3,000 students annually. Topics in the last couple of years have continued the rich mix of systematic and regional, human, physical, and technical topics. Given care and good understanding of what you wish to present, geography can come alive even for the youngest students. Even by the most stringent standards, this decade–long effort to reach out to young people in the community and inform them of the nature of today's geography has been an unqualified success.

Some of the factors that have contributed to this success include: department commitment; no slackers; encourage senior faculty to set examples by being among the first volunteers; start your advertising well in advance; go through appropriate channels—the superintendent of the school district, school principals, and then to the teachers (via an invitation posted by the principal.) Do not underestimate students even in the early elementary classes, don't speak down to them, and give them every opportunity to ask questions (i.e., don't just give a lecture, make your presentation active). Use geographic media—globes, maps, visualizations, computer software, slides, videos—whatever it takes to capture interest. But don't use too many of them. Give the students a particular task during your presentation. Send some background information on your talk to the teacher, and suggest a few projects that the class could do as a follow up to your presentation.

Hold a pizza party at the end of each Geography Awareness Week, during which the presenters can talk informally to each other about their classroom experiences. These procedures have proven successful for our program. It also doesn't hurt to notify local television and radio stations and the local press about National Geography Week activities, together with an indication of which schools are being visited, when the visits will occur, and what topics are to be discussed.

There will be problems. First, some school principals will file your request for cooperation in the wastebasket without notifying their teachers. We found that even when this happened year after year, the teaching network spread the news of what was going on and individual teachers called the departmental contact and asked for visitations. The district superintendents may not be enamored of the idea because of a fear that such presentations may set back teachers' efforts to complete the set curriculum in a timely way. At times presenters have turned up to schools expecting to find high–tech equipment for their presentation, only to be seriously disappointed. Always take a fallback presentation module. The usual setbacks happen—missed buses, cars with flat tires, geographers getting lost on the way to the school, going to the wrong school, turning up at the wrong time, and on and on. To deal with these situations, carry a contact number and a cell phone. You will probably find that it is harder to get invitations to present at middle schools and high schools than it is to present at elementary schools. This is simply because the elementary school teacher has more flexibility in the way they deal with their classes. Be flexible—offer to come back and visit at some other time of the year when your topic fits more neatly into a curriculum slot.

Changing the image of geography and increasing the awareness of the nation's population regarding geography is, I believe, the first step in what should be a massive attack on geographic illiteracy. Part of the attack, however, must be to show K-12 students that the profession of geography is an exciting one and has long-term employment potential. You can turn the kids on all you want, but if there is no clear link between what you are talking about and an employable future, that excitement generated by your presentation will soon fade.

I don't think anyone knows how many students and other people are being reached during Geography Awareness Week. My guess would be that it is only about 10% of what should be reached. So, I throw out a challenge to you as members of the AAG. During that week of awareness, get out to schools and the local community and represent both the discipline and your particular areas of interest. In the August *Newsletter* column I stressed the need to use our varied human resources in a more effective way. Geography Awareness Week is one event in which the full range of resources of the profession can be used. Retired geographers can talk to schools about their areas of interest just as can current faculty, students, and the various staff components of the department. All you have to do is organize their participation.

Let's pull out all the stops and really start to make people understand what an imaginative, productive, and intriguing discipline it is that we all practice.

I can answer questions people may have about how to begin this type of project.

ALEXANDER B. MURPHY: *An Imperative (And an Opportunity) for Geographic Education*
September 2003
AT THE END OF A TALK I recently gave to a high school Advanced Placement Geography class, one of the students asked me to identify my favorite travel destination. It was an impossible question to answer, of

course, but I thought for a moment and replied, "the place I haven't been to yet." I suppose the answer wasn't quite what the student was looking for, but I just returned from one such place, and the experience has left me thinking that I gave the student exactly the right answer. The place I visited was Iran, and the experience was extraordinary.

I went to Iran to address the Second International Congress of Geographers of the Islamic World. My participation in the Congress also gave me the opportunity to make presentations at two of Tehran's universities, to discuss contemporary issues with the media, and to see some of the country. My experiences and impressions could fill a year's worth of presidential columns, but one reaction to my trip dominates all the others: the extraordinary importance of challenging the stereotypes people hold about one another.

In making this point I know that I risk sounding incredibly trite. Yet on both ends of my trip I was struck by the generalized view that each side has of the other. When two places are separated by half a globe some stereotyping is probably inevitable, but the magnitude of the misunderstandings between the United States and Iran has reached dangerous levels.

A very specific set of circumstances has, of course, fed into the stereotypes that divide the two countries. For many Americans, Iran conjures up images of hostages being held captive in the American Embassy in Tehran, of dissidents disappearing in the wake of the Islamic Revolution, or of the *fatwa* issued against the author Salman Rushdie. For many Iranians, America brings to mind CIA efforts to intervene in local politics, US military aid to Iraq during the 1980s, or the worst excesses of an unrestrained pop culture.

These impressions are all rooted in actual events, but they obscure a much more complicated—and ultimately more hopeful—reality. Within Iran, for example, there is significant opposition to some of the repressive aspects of the current regime among a population that is quite divided about where the country should be going. Moreover, the reality of Iran utterly belies the simplified notions of "the Islamic World" that are often bandied about. Not only does Iran see itself in sharp contrast to its neighbors because of its distinct history, culture, and contemporary experience; the country itself encompasses a wide spectrum of cultural practices and political orientations.

All of this suggests a great challenge for geographic education: to expose the complex geographical realities that lie behind—and frequently undercut—the dominant geopolitical generalizations of the day. The challenge is as critically important as it is vast. Whatever position one may take on recent developments in North Africa or Southwest Asia, we cannot even begin to have an intelligent discussion about them if terms such as "the Islamic World" or "the Middle East" are deployed in a highly caricaturized way. My visit to Iran impressed upon me the need not just to teach our students about the difference between Persians and Arabs, or to make sure they know that Iran is the state with the largest Shiite population. There is an even greater need to deconstruct the notion that Shiites are necessarily repressive fanatics, that Iranians have a monolithic view on the issues of the day, or that Iranians are victims of a backward culture.

Much has been written in recent years about the ways in which geography has served the interests of military conquest, colonialism, and economic exploitation. Yet geography can be an equally powerful tool in the effort to forge a more peaceful, stable, and environmentally sustainable world. As the stakes rise in confrontations over people and the environment, so also the importance of geographic education rises. And there can be few causes more worthy than combating the dehumanizing stereotypes that so often seem to be the currency of the day.

One experience I had in Iran drove home to me the importance of this project. During the course of my visit I was asked to appear as a guest on a current-events television program that draws a wide audience. The topic that day was the changing geopolitical scene in the wake of the "War on Terrorism." My aim was to discuss, in as frank and honest terms as I could, aspects of the emerging geopolitical picture in the region. I tried to make some carefully crafted points, but it soon became clear to me that my specific comments were far less important than the overall impression I conveyed. By coming across as an American who understood and appreciated something of Iran, who did not buy into the simplifications that dominate news coverage on either side, and who was a clone of neither Donald Rumsfeld nor Eminem, I was in a position to challenge one set of stereotypes. As geography educators, it is imperative that we challenge the stereotypes that run in the other direction. Whether trite or not, such endeavors carry with them hope for a better world.

❊ ❊ ❊

Chapter 4

CAREER PREPARATION AND ALTERNATIVE
PROFESSIONAL PATHS

❋ ❋ ❋

Preparation for a career in geography dominated the thoughts of some of the AAG Presidents who expressed concern about the appropriate intellectual content and skills required for a career in contemporary geography. At the same time, other columns were devoted to the diverse pathways that professional geographers have pursued, highlighting the role of non-academic geographers and the successes they have achieved.

❋ ❋ ❋

RONALD ABLER
August 1985

A SERENDIPITOUS BENEFIT of my current sojourn in Washington has been the opportunity to meet the local geographical community. The metropolitan area contains one of the largest concentrations of geographers in the world. Some are academics, but most work in government or private industry where they practice geography in much the same way that a physician practices medicine. Practitioners comprise a fifth of the Association's members. They practice geography in government, in private industry, as self-employed individuals, in non-profit organizations, in research centers, and in the military.

The work of practitioners is critical to geography's welfare. People outside the academic world learn of geography largely through the work of practitioners or applied geographers. Government officials and business and industry leaders may occasionally encounter a specific item of published geographical research, but any day-to-day awareness of what kind of work geographers do and can do most likely comes from a practicing geographer.

Many practitioners believe—correctly, from my observations—that academic geographers undervalue their contributions in two ways.

First, virtually any practitioner who has solicited recommendations from an academic department (especially a doctoral department) can recall comments along the lines of "Student X is not capable of pursuing an academic career but is well qualified for your kind of work."

Besides being gratuitously insulting, such comments reveal a shortsighted attitude. Academics do the profession a disservice when they discourage talented, energetic students from pursuing careers as practitioners. Geography needs practitioners to advance the discipline in quarters where it is even less well known than within the academic community, and that task is worthy of a share of the best talent we can muster.

Second, practitioners feel underrepresented within the Association, where they have not been as prominent as they should be on committees, the editorial boards of journals, the AAG Council, and among elected officers. Representation has improved recently, and I hope AAG members will continue to be more receptive to greater participation by practitioners in professional affairs.

A healthy discipline needs a yeasty balance of theory, method, and applications. Practice is where theory and methods meet the acid test of applicability to the world's workaday problems. The AAG is primarily a scholarly organization devoted to advancing theory and method, and it is likely to stay that way. The majority of its members are academic geographers, who have more time to devote to AAG affairs and greater incentives to do so. But the AAG is not exclusively an academic organization, nor should it be. That was determined in 1948 when the ASPG and the old AAG agreed to combine forces in a single association.

Geography is too small and too vulnerable to afford factions or factionalism. Advancing the discipline's interests requires our unified talents and energies. A broader appreciation of the invaluable contributions practitioners make and greater efforts to involve them in AAG affairs can only strengthen our common cause.

Robert Kates
July 1993

A LETTER ARRIVES from the newsletter editor with a "friendly reminder that your first President's Column is right around the corner" and suddenly I know that my apprenticeship is over. Being vice-president of the AAG is a time for learning under the friendly tutelage of the AAG staff, officers, and council. Now it's my turn to follow Tom Wilbank's example, whose thoughtful, provocative use of this column sets a standard that I can only hope to emulate.

I am now an independent scholar, at least that is the way I designated myself on my name badge in Atlanta. It's not that I lack titles and affiliations. I am University Professor Emeritus at Brown University as well as Adjunct Professor of World Hunger, Affiliate Professor of Geography at Clark University, and Faculty Associate at the College of the Atlantic. I am also one of four executive editors of *Environment* magazine and also serve as co-chairman, along with Akin Mabogunje, of the international initiative entitled Overcoming Hunger in the 1990s. But I wanted to be an independent scholar and pension funds and social security make it possible to sustain oneself. And mail, e-mail, fax machines, and travel make possible continuous access to my real affiliation, the invisible college of friends and collaborators.

Nor is this the first time that I have been an independent scholar, as there was a decade-long period in the mid '70s and '80s when I resolved some of my conflict between teaching and research by becoming a half-time independent scholar. I even acquired membership in the Academy of Independent Scholars. Unfortunately, that Boulder-based initiative of Kenneth Boulding and Larry Senesh did not survive funding difficulties and the absence then of a communications network such as Internet. But it did establish the notion that scholarship, even great scholarship, can come from settings other than the conventional academic ones and at times late in the life cycle.

My all-time favorite independent scholar is Ester Boserup, who created a major theoretical alternative to Malthusian resource theory as well as initiated an enormously fruitful line of study of women's roles in development. But her example also raises an important question. While I am not aware of her life choices, women certainly more than men have independent scholarship thrust upon them. Involuntary independent scholars also include new PhD's without academic appointments, two-profession couples forced to be content with a single available post, late bloomers and vocation changers repeatedly passed over, mavericks without niches tolerant of their difference, and even reluctant retirees. Indeed for quite a few geographers, being an independent scholar is not a desired status to be worked towards, but a type of second-class professional citizenship that has been forced upon them.

Are there some things we could do for independent scholars, enthusiastic or otherwise, to make their lives easier and more productive? A good way to start might be to consider the thoughtful advice offered by Catherine Reed, an independent scholar ecologist urging jobless ecologists not to abandon research. She gives practical advice under the following headings:

- Find at least one institution and one individual to facilitate your work.
- Develop a long-term research project and keep working on it.
- Develop a cheap project. Do your research locally and make it applicable.
- Look for alternative funding sources, including local agencies and foundations whose actions may be influenced by your work.
- Develop alternative labor sources, especially for summer field projects.
- Maintain your graduate school contacts.
- Model your life on the artist's life.
- Keep your expectations low.
- Keep your self-confidence high.

Read the piece in its entirety in *The Scientist,* May 3 1993, page 11, and then consider the implications not only for the independent scholar but for potential hosts. Does your department or institution create non-remunerative, but still essential, affiliations for independent scholars that provide a title, library access, computer links, e-mail, and the required institutional base for grants to be awarded? Can we make regional meetings hospitable places for our independent scholars to whom the costs and time away for national meeting may be prohibitive? Do we go out of our way to include independent scholars on our mailing lists for seminars and speakers, invite them to talk to our classes, or offer them critical review of their work or other professional mentoring? And of course, to my fellow independent scholars out there, write and suggest what would be helpful from your experience.

LAWRENCE A. BROWN: *Reaching In*
March 1997

My last column focused on reaching outside our profession. Here, I focus on talking to one another, a preamble to this year's Presidential Plenary Session—Change, Continuity, and Discourse: The Challenge To and From Geography.

Each of us needs to maintain integrity as professionals, yet be open to change as new venues of scholarly discourse are explored and prove useful. There is a progression to this process—the dialectic of science.

As a graduate student at Northwestern and young professional at Iowa and Ohio State, I fell into the quantitative revolution. Contestation involved idiographic versus nomothetic approaches, fieldwork and knowing one's study area versus more armchair approaches, the imperative of statistics, mathematical modeling, spatial laws, and positivist reasoning. At one time, articles featuring quantitative analysis were excluded from the *Annals*, later relegated to Technical Notes. Geographers of my ilk embraced Regional Science, published in RSA journals, established our own *Geographical Analysis*, and many either eschewed or participated minimally in the AAG. Today the debate has come full circle. Quantitative analysis started as antithesis. Later, with synthesis, it was merged into our kitbag of tools such that virtually all departments offer quant courses. But now, spatial analysis is for many a bête noire, the thesis that must be confronted, the norm to be altered. A familiar, dialectic of science scenario!

In earlier columns on McBryde and Sauer, I noted that the quantitative revolution had strong roots in earlier research. This is illustrated by Hagerstrand, whose work appeared both in Wagner-Mikesell's *Readings in Cultural Geography* (1962) and Berry-Marble's *Spatial Analysis* (1968). Change, Continuity, and hopefully, Discourse!

The discourse of scientific inquiry is personally important. Perhaps, however, to paraphrase a Pogo character (which??) ."We have seen the troglodyte and he is us!"

I've witnessed ongoing discourse on how research should be done, the subjectivity of objectivity, and the real meaning (or lack thereof) of research. It's important to critique existing modes of scholarship, an integral element of the dialectic of science. It's important that geography's focus on deterministic views of human behavior gave way to concepts of individual choice, constrained choice, structural impacts, and contingent outcomes reflecting in interaction of broad forces with individual people, places, etc. Such shifts occur via the *isms* of scholarly inquiry—the quantitative revolution, behavioral geography, Marxism, realism, structuration, social theory, post-modernism, and new geographies (regional, cultural). Is each conversation overly protracted? Often. Is there over-marginalization of established practices and related individuals; and over- lionizaton of new age professionals, temporarily or permanently? Yes. Having been on both ends of the continuum, I understand this is the way of academic discourse. Nevertheless, while intellectual differentiation is critical at the beginning of change, it eventually becomes dysfunctional. Ultimately, then, the issue is Change, Continuity—and Discourse.

Personally, a fundamental precept is, "It's the research question, stupid," and the litmus test of an *ism*, the bottom-line, is its effect on doing research. I've benefited from the many *ism*-debates. An exemplar from realism is the insight that outcomes are contingent. I embraced structural perspectives on my own, but that became more legitimized through discourse related to political-economy paradigms. I continue to bristle over my longstanding contention that, in terms of doing research, realism and Chorley's 1960s statement of open system thinking say the same thing. And concerning the subjectivity and social creation of meaning, objects, ideas—this is nothing new to readers of existentialist literature. Nevertheless, I understand the argument needs to be put forth repeatedly, different strokes for different folks. But also, those putting forth the argument need to recognize ties to the past, the literal and figurative antecedents of current ideas—and that talking about and doing research are dramatically different endeavors. Change, Continuity, and Discourse.

Terminology has a role. Bewildered by the word *fractal*, I boldly asked its meaning, and found an equivalency with *isomorphism*, from earlier research epochs. Connecting these terms might omit nuances and be contentious for some, but invariably such connections have validity and need confirmation. I've railed against the terms *capitalist*, *pre-capitalist*, and *capitalist penetration*, leaving gigantic circles on student dissertations.

I now concede the battle, and even use these terms myself (but not yet *hermeneutics*!). Nevertheless, while locales are penetrated by capitalism in varying degrees and ways, there is no such thing in today's world as *pre-capitalism* or a lack of capitalism. I've seen research that treats sub-populations separately under the guise of *deconstructionism*; yet the endeavor, perhaps mislabeled, is a long-standing social science procedure. *Praxis*, according to Webster, means practice; simple, yet the word uttered in a scholarly setting intimidates many. Terminology can separate or unite. Change, Continuity, and Discourse.

Relevant to this discussion is a talk in response to the motif of this year's SWAAG meeting, "Celebrating Fieldwork." I sought to demonstrate that fieldwork encompasses a broad range of activities, the only common denominator of which is being there in some form—an inclusive, not exclusive definition of fieldwork. Another theme is the researcher's need to trust herself, whether insight contradicts or reinforces accepted knowledge—if a scientific construct does not ring true to one's experience, perception, feelings, it (expert knowledge) should be immediately open to question. This empowerment follows because, in human geography, researchers are a part of the phenomena being studied; their own experiences. Insight, instinct, perception, vision is a critical ingredient. So living is itself an important component of fieldwork, and if one has ventured forth in the world with an inquisitive, experiential, open, listening perspective, that's an essential ingredient of our scientific endeavor. Finally, in many aspects of human geography, particularly area studies, "fieldwork" is a *sine qua non* of quality research, a rite of passage that signifies one as a true geographer, a mantra that provides membership in an exclusive club. In the end, however, fieldwork is simply another research tool—and not any more sacred than, say, regression analysis. We know that the selection of a research tool(s) should be guided by the substantive question at hand, a means of gaining suitably intimate knowledge of the phenomena being studied (place, process, social-cultural group, etc.), an understanding of ground-level reality. Said another way, the substantive dog should wave the methodological tail. But too often we find geographers behaving otherwise—in a manner such that the methodological tail waves the substantive dog. It's the research question, stupid!

So, while perhaps bombastic and the troglodyte among us, I want to be educated, understand better, broaden and deepen my abilities. Change, Continuity, and Discourse: The Challenge to and from Geography will be a step in that direction, made by persons far more knowledgeable than I – Susan Hanson, Victoria Lawson, and Diana Liverman. Join me in listening to them.

PATRICIA GOBER: *Rethinking PhD Education in Geography*
September 1997

Scientists are rightly concerned about the dearth of permanent jobs for new PhDs in physics, biology, chemistry, and geology. Many young scientists are disappointed and frustrated because they spent their youth preparing for jobs in science that no longer exist. Post-doctoral positions, once stepping-stones to permanent faculty positions, have become careers in themselves. After graduation, many scientists find themselves part of a growing cadre of intellectual migrant workers moving from college to college from one part-time or temporary position to another. After considerable soul-searching about the advisability of creating a two-tiered faculty, this year my own department advertised for a temporary instructor to teach world geography exclusively. When we were swamped with applications from excellent candidates from some of the most prestigious PhD programs in the nation, I concluded that geography is not immune to the forces that have so profoundly affected the job market for other scientists.

A recent report of the Committee on Science, Engineering, and Public Policy, a joint committee of the National Academy of Sciences, the National Academy of Engineering, and the Institute of Medicine, recommends that graduate programs in science should respond to the limited job market for academic positions by providing a broader exposure to experiences desired by both academic and nonacademic employers. Graduate education should prepare students for an increasingly interdisciplinary, collaborative, and global job market and should not be viewed only as an intensive research experience.

It is my impression that, with some exceptions, PhD programs in geography continue to follow the traditional model of graduate education, one that emphasizes a highly narrow research experience. PhD programs are geared toward perpetuating themselves by producing graduates who are well trained for the highly-specialized research and teaching requirements of other PhD programs in geography.

The rub is that not many of our new PhD graduates in geography will hold positions in PhD departments. Indeed, many do not aspire to such positions. To better understand the current job market for new PhDs in geography, I examined the job listings in *Jobs in Geography (JIG)* for a 12-month period from August 1996 through July 1997. Out of the 201 positions advertised, only forty-three involved faculty appointments in freestanding PhD programs in geography. Thus, only about 20 percent of our new PhDs are likely to hold the types of jobs for which they have been prepared. Others will work in undergraduate and master's institutions; in joint departments where geography is combined with geology, planning, or anthropology; in interdisciplinary programs; or outside the academy altogether in government and business.

Job listings in *JIG* also reveal the importance of an interdisciplinary perspective to finding a job in today's job market. More than 35 percent of jobs are in interdisciplinary programs focused on women, Chicanos, area studies, global studies, environmental issues, and policy analysis, and another 20 percent are in joint departments where geographers interact daily with colleagues from related fields. Even advertisements for jobs in geography programs stress the need to collaborate with colleagues outside the department.

PhD programs that continue to stress narrowly-defined, discipline-specific research experiences are badly out of touch with the realities of the current labor market for PhDs in geography. While graduate education in the US should retain the features that have made it a model for the rest of the world, including an original research experience, more can be done to impart a broader range of skills. We need to ask ourselves how effectively our current PhD programs serve the students who face today's job market and will face the one 10 years down the road. How many interdisciplinary experiences do we provide? How much emphasis do we place on teaching and how much support and mentoring do PhD students receive when they do teach? Do we expose students to the range of teaching environments in which they actually will find themselves? Do they obtain administrative experience? Do we offer or require internships with the private sector and government as part of the PhD experience? Are we open to dissertations that tackle real-world problems, the kind of research that is valued in business and government? When I was a graduate student at Ohio State, I was lucky to have had the opportunity to work part time at Battelle Memorial Institute in Columbus doing applied research. While I ultimately chose to pursue an academic career, my Battelle experience helped me to make a good career choice and served me well in relating to the majority of our graduate and undergraduate students who seek professional, not academic, careers.

Concern about the narrowness of traditional doctoral education has become a hot topic in higher education circles. The Pew Charitable Trusts recently funded a program called Preparing Future Faculty that is designed to expose PhD students to a wide range of instructional settings, from community colleges and private liberal arts colleges to comprehensive universities. Arizona State University, along with the University of Washington and the University of Minnesota, were among the first recipients of Pew grants to develop prototype programs. The ASU program, enrolling some seventy students this year, is two years in length. The first year introduces students to faculty roles at a variety of institutions, legal issues in higher education, distance and cooperative learning, preparing for the job market, and the future of institutions of higher learning. In the second year, program participants become apprentices in partner institutions for team teaching, developing new courses (one PhD student developed a new statistics course for a local community college), and shadowing faculty members and administrators.

I doubt that many PhD departments have the time, energy, or expertise to develop their own programs on this scale. Still, we can look carefully at our curricula to see whether, in addition to a thorough grounding in the fundamentals of their subfields, students obtain a broad familiarity with other disciplines, whether they gain experience in various types of classroom situations, and whether they gain work experience in nonacademic settings. I'm talking about a new kind of PhD, one that emphasizes adaptability and versatility in addition to technical and specialized expertise. The combination of these traits would give our graduates an invaluable edge in today's and tomorrow's job market, and it would give geography a way to contribute in a much broader range of problem-solving and decision-making settings.

WILL GRAF: *Not Clueless, Just Skill-less*
January 1999

ONE OF THE MOST IMPORTANT THINGS geographers offer to society, government, and other sciences is our perspective on the world. We emphasize space and place, location, patterns, networks, systems with locational characteristics, and the spatial perceptions and behavior of people. We contribute a better understanding of the present world and improved predictions for a future one. We offer clues that other disciplines overlook. In order to be effective teachers, researchers, and public servants, however, we need to offer more than clues. We must also offer skills to actuate our knowledge, and it is here that we need to make substantial improvements. In this column I outline the lack of skills and some potential solutions.

While physical and human geographers often lack important analytical skills, technical specialists sometimes have the opposite problem: all skill and no insight. Physical geographers deal with physical, chemical, and biological systems from a spatial perspective. They are not geologists, atmospheric physicists, chemists, or biologists, but rather spatial specialists who complement other scientists. Emphasis on the geographic characteristics of environmental systems does not excuse physical geographers from knowing something about the fundamentals of their chosen specialties. Geomorphologists, for example, must understand basic physics and chemistry to understand the processes and forms of the earth's surface, while hydrologists and climatologists need a clear appreciation of fluid dynamics, a subset of physics. Biogeographers are not likely to contribute effectively to their field without knowledge of the fundamentals of biology. As a discipline we train too many physical geographers with too little of this cognate knowledge. Students emerge from many of our undergraduate and graduate programs without even a high-school level of knowledge of physics and chemistry. Even worse, they lack mathematical skills commensurate to introductory calculus.

Human geographers deal with the spatial aspects of society. They are not sociologists, urban planners, or economists, yet they often interact with (and compete with) specialists from these disciplines. The reason geographers deal with questions of society is that they emphasize the roles of space and place in social and economic processes, but like their physical geographer colleagues, they require knowledge of cognate disciplines. For many human geographers, interviews and surveys are a critical part of the data collection process, but we rarely encourage our students to obtain formal training in survey research. At the very least such training allows our students to avoid costly mistakes, and in the best case it would enhance their research products. Economic geographers without training in econometrics are not likely to convince potential consumers that their insights into regional and network processes are useful. Human geographers of a more qualitative bent require formal training in landscape interpretation, ethnography, case study approaches, and narrative interpretation.

The technical specialists in geography have an abundance of skills, but often make a reverse mistake; they shun theoretical aspects of geography. They become adept at generating images or maps, but lack the critical thinking skills necessary to stretch geographic technology to its productive limit. In part, this lack of theoretical background explains the present conundrum in GIS: it is a great way to make pretty maps, but we greatly underutilize the potential of GIS for analysis. This critical underachievement is becoming a severe limitation for expansion of the field. A friend recently defined the problem: coming to geography to learn only GIS is like taking a degree in English to become a typist.

The solution to these issues is in the hands of the education community. Faculty are responsible for providing sound advice to students about what they will need to succeed in a competitive academic, governmental, and business world. Physical geographers need at least introductory knowledge of basic physics, chemistry, and mathematics. Human geographers need at least introductory knowledge of mathematics, statistics, survey research, and the fundamental ideas in related fields. GIS experts need to delve more deeply into the underpinnings of spatial analysis. Students need to protect their own interests, and if their geography programs are weak in the skills department, they should demand more suitable training.

In addition, there are a series of skills that our graduates tell us are often the most important things they learned in their university experiences: data-related proficiencies and communications skills. Natural and social scientists know the world partly through their data, and we cannot afford to neglect the training needed to obtain and make the greatest use of available data. Database management is becoming a critical tool as more public money is invested in collating the data gathered by governmental agencies. These resources, with much of their data connected to locational attributes, represent part of the route to explanation and prediction for many of us. They also represent critical inputs to public policy debates, regulatory efforts, and environmental or economic management. Although some of us may speculate that quantitative analysis is a historical artifact, quantitative methods remain a central feature of science, business, and government. Effective use of GIS is impossible without knowledge of the quantitative underpinnings of such systems.

Finally, few of us are blessed with the natural skills of effective public speaking and writing, but most of us can learn. We need to spend time with our students developing these skills or send them to others who can teach them. No matter how insightful and thoughtful our students are, failure to communicate ideas and results will doom them and their contributions to obscurity.

We need to maintain our focus on the basic concepts of geography, but we must also not lose sight of the importance of the fundamental skills our students need. We need to require that students who complete our degrees are well skilled because if they are not, they are less likely to succeed and they will represent us poorly. We need to advise students about where to obtain the skills if they do not have them or if we cannot teach them. The workplace for our students rewards those who can combine basic geographic knowledge with technical expertise, but the future is dim for those who are skill-less. If students are not convinced about the need to invest the necessary human capital to obtain the skills, invite them to consult two sources of reliable information: past graduates and current job announcements.

JANICE MONK: *Valuing Service*
July 2001

As I ASSUME the challenges and opportunities of the Presidency of the Association, I have been thinking about what motivates geographers to undertake professional service, the ways in which such work is rewarded and rewarding, and how we foster commitment to service among geographers at all stages of their careers. So many times I have heard young professionals advised to avoid service because it "doesn't count." Some department chairs believe they are obligated to protect junior faculty from service, assigning just enough so that the category doesn't show up empty at tenure and promotion time. I know less about the rewards for professional service in business and governmental agencies, though certainly some companies support community service.

But service is critical for sustaining much of our professional work—someone has to review manuscripts and grant proposals, edit journals and newsletters, write letters of recommendation, manage listservs, organize conferences, sponsor student organizations, review departments, lead the Specialty Groups, serve on the Association's Committees—the list goes on. Locally, building relationships across departments and with administrators, fostering student organizations, and connecting with schools are critical to program health and survival. To show the value of geography to society, we need to communicate our work through the media and to people in public and private agencies. Even though we may protest about the seemingly escalating demands of our work, we know that much of it is essential to sustaining and advancing our profession.

How do we create a climate that encourages and supports professional service in our educational programs, in our places of work, and in the Association? Looking to the future, two years ago I interviewed three graduate students I knew to be active service contributors—Kimi Eisele, then at the University of Arizona, who was instrumental in founding *You Are Here: The Journal of Creative Geography,* a highly professional publication edited by graduate students; Kim Elmore of the University of North Carolina, Chapel Hill, a cofounder of the organization Supporting Women in Geography (SWIG); and Susan Mains, then at the University of Kentucky, who managed the GeoFem and LeftGeog listservs for four years. They identified significant rewards and learning from their experiences, and also spoke of ways in which they had or had not been supported.

Kimi Eisele thought she became a better leader as she struggled to get ideas out of her head and enough into the hands of others so they felt some sense of ownership. She learned "a lot about management, opportunities, organization, grant writing, public relations, and how to channel creative activity" and "how to explain the bigness of geography to non-geographers." Kim Elmore reported a sense of accomplishment in collective activity, an opportunity to meet other like-minded and active geographers, and to generate interest in her department, on campus, and beyond. Susan Mains emphasized the importance of connectedness between her service and her learning:

> *"It has played a crucial role in making me feel part of geography. It has taught me that diplomacy is important, but so too is the ability to be assertive and keep open, as much as possible, spaces in which a variety of geographic perspectives can be expressed."*

These young geographers clearly have visions that extend beyond "getting through" and establishing personal reputations. Their service activities have nevertheless provided them with opportunities not always gained through class work and research—to develop leadership, widen their horizons, and learn and practice an array of professional skills. There have also been the personal rewards of developing friendships.

How has their work been viewed locally? They report peer and faculty encouragement, but also indifference and some implicit, if not explicit discouragement. The encouragement came from individual mentors, fellow students, clerical staff, administrators, and structural opportunities within the universities. But they also encountered indifference and discouraging voices that conveyed the low value placed on service and the limited resources available for such work. Ironically, the drive to professionalize students can serve as an obstacle to undertaking service, as students are prompted to publish in order to better their prospects for employment. How can we promote a more balanced view?

From faculty, I hear increasing distress about demands for committee and service work as institutions are pressed for accountability. The expansion of part-time positions in academia places more pressures on those in full-time appointments to do the work of sustaining institutions. The situation is likely similar for geographers in business and government. How can we promote a vision that engagement with the collective is critical for individuals, institutions, the discipline, and society? To work towards that end, we need to identify opportunities for service that are meaningful, rewarding and rewarded.

In the Association, we will require more, not less, service if we are to build bridges across fields, sectors, with other disciplines, and to heighten public awareness of our contributions. Each year we honor colleagues with Distinguished Service Awards. Some, but not all, of our specialty groups make similar awards. Mentoring panels are periodically offered at national meetings. What other avenues should we explore and promote to foster and support service?

JANICE MONK: *Changing Careers*
March 2002

SEVERAL RECENT CONVERSATIONS prompt me to write this month about the nature of careers in geography, some ways in which they have been changing, and how geographers experience their careers. We have already begun to see the retirements of the substantial group who entered the profession in the late 1960s when the job market was booming. They could expect relatively secure academic appointments with predictable advancement in rank. Today's entrants may have wider choices but also more limited opportunities. On the one hand, technical developments have opened new prospects beyond academia. On the other, the conjunction of a business orientation and budget crisis in higher education have increasingly pushed scholars into insecure, part-time, or temporary appointments.

Other changes are evident. Whereas the retiring group predominantly includes white men, today there are increasing numbers of women. Often they have different needs and expectations about combining their personal and professional lives than their predecessors. Issues are especially complex for dual-career couples. How can and should institutions accommodate them? We have gone from anti-nepotism policies to struggling with the politics and funding of partner placement. What is reasonable for professionals to expect? What prices are paid in personal lives in order to sustain professional work?

The needs and desires of individuals over their careers and the priorities of institutions are not always congruent. Should employers promote early retirement packages in order to open up new positions and make budgetary savings, or encourage people to stay on rather than lose faculty lines? How do we balance personal and institutional needs? How do we prepare and sustain geographers for fulfilling careers across the life span?

Answers to these questions are neither universal nor all within the purview of the profession. In this column I'd like to draw attention to some of the programs that are addressing career development, pose some additional needs, and invite you to suggest steps that might be taken locally and through the AAG to enhance the ways geographers experience their careers.

Most attention focuses on preparation and early career stages. One such effort is the Commission on College Geography II project that addresses the development of teaching skills among new faculty and advanced graduate students. It recognized that traditional doctoral programs often don't systematically include this component.

The prestigious individual CAREER Awards offered by NSF to foster leadership and promote the integration of research and teaching are another example. Four geographers—Linda Barrett (University of Akron), Gavin Bridge (University of Oklahoma), Meghan Cope (SUNY–Buffalo), and Matt Sparke (University of Washington) currently hold these awards. They have exciting agendas that mesh with the missions of their institutions by bringing together cutting–edge research in geography and educational projects designed to serve those they are researching, as well as local communities. Among the activities of these scholar-teachers are creating internet-based, student-centered learning activities on environmental themes for undergraduates with little science background; teaching in Guyana and Peru in conjunction with field research there on land use and land cover changes resulting from mining; bringing undergraduates into low-income and minority schools to investigate children's geographies and draw on this research to create teaching materials; and supporting an ambassador outreach program to high schools in minority neighborhoods as well as service learning options in non-governmental organizations for graduate students with the aim of bringing students into better touch with their geographical positions in a more globally integrated world.

In contrast to the support these geographers are receiving is the situation of young itinerant professionals who take on a series of temporary jobs as they pursue elusive tenure track positions. They face seemingly endless new preparations and personal upheavals. One recently told me of teaching sixteen different courses over the several years it took him to locate a position with permanent prospects. Another, describing a course she was assigned to teach admitted, "What the hell do I know about that!" Is there anything the profession can do to alleviate the stresses of these situations?

The recent survey by the National Association of Graduate and Professional Students (*http://survey.nagps.org*) draws attention to aspects of graduate education that might be strengthened to give new professionals more options. Geographers who participated in the survey were least satisfied with the limited extent to which their doctoral programs encouraged them to explore a broad range of careers beyond academia or to prepare them for those options.

Early career support is certainly important, but what about mid-career development, other than through sabbatical leaves? What kind of networking, training, or other support would be helpful to those who might wish to move across sectors, permanently or temporarily—to widen their horizons, seek new challenges, or share research experience with practitioners and vice versa? To think about career development over the life span, including the transition to retirement and beyond, Lydia Pulsipher is bringing together a discussion group of women and men geographers at different career stages. I look forward to co-hosting this conversation and hope we can articulate creative approaches to share with others.

For those beyond formal employment, the Retired Geographers Organization (Charles Bussing, President) seeks ways to draw on members' talents and energies. Some have led highly successful overseas study tours. Others are pursuing web-based or writing projects to reach popular audiences. What other roles are possible and fulfilling for the so-called "retired?"

As we look to the future of geography and the AAG, we need to be thinking about careers in more complex and diverse ways than before. I would like to learn of your concerns and suggestions for collective action.

M. DUANE NELLIS:
Crossing Disciplinary Boundaries: Missionary Geographers
November 2002

MANY OF US have our homes in an academic geography department but some geographers have found niches in academic programs outside geography, in private industry, or in government agencies. Many of these individuals remain active in the AAG. Some do not, even though they actively promote the discipline of geography, because they fail to see the value of their work to the Association or the benefits the Association offers them. I worry that some in the discipline may not recognize the talent and perspective that those who work outside academic geography departments bring to the discipline. We need to embrace all practicing geographers regardless of their affiliations, and work together to promote geography's presence in all arenas.

As missionaries for the discipline, geographers outside the traditional academic setting exhibit to their colleagues from other fields the concepts, techniques, and unique perspectives, approaches, and solutions that geographers offer. People like Richard Marston, the Sun Distinguished Professor of Geology at Oklahoma State University, and AAG Past President Janice Monk, Director of the Southwest Institute for Research of Women at the University of Arizona, are two examples of those carrying out such geography missionary work in academic programs outside geography. These are but two examples of the many geographers the AAG should actively seek to keep involved in the Association and in the discipline in general. They may have allegiance to other professional organizations or be expected to become active in other associations, but that does not preclude continued loyalty to the discipline that trained them and to remaining active in the AAG.

One of the themes in my AAG Presidential candidacy statement was the importance of supporting the Association's Strategic Initiatives, including building partnership with geographers more broadly. Professional geographers outside traditional academic geography programs are truly missionaries—they provide others with a unique perspective on geography, and they raise the visibility of and respect for the breadth of spatial perspectives that the discipline provides. The Strategic Initiatives are designed to foster greater coordination and integration of public, private, and university efforts to advance geographic research, education, and outreach. They can also provide an important forum for fostering dialogue between these constituencies.

There is also a whole cadre of professional geographers outside academic geography who are advancing the field across a broad spectrum of the government and private sector. People like Kamlesh Lulla, Chief of the Earth Sciences Branch at NASA Johnson Space Center; Bob Marx, head of the Geography Division at the US Census Bureau; and Barbara Ryan and John Kelmelis, USGS Senior Scientists, are examples of leaders in the government sector. These are people who have been important, longtime members of the AAG, who are committed to furthering geography on numerous fronts. There are hundreds of other geographers in this sector who can help advance the Association's work. Should we be sponsoring special forums within our national meetings that would appeal to this broader constituency and enhance the opportunities for interaction and collaboration between this group and academic geographers?

There is also a rapidly growing group of missionary geographers in the private sector, who provide support for the discipline, but who should be further encouraged to partner with academic geographers and the AAG to support innovative research and internship opportunities, which would strengthen the discipline. C.J. Cote, who is in charge of international programs at ESRI; Marge Elliot, President of Geo Insight International; and Juha Uitto at the World Bank, are examples of these important links to the private sector.

How can we strengthen our ties to this important group of geographers? I suggest we: 1) appoint them to serve as adjunct members in geography departments; 2) ask them to serve on AAG committees or run for AAG office; 3) encourage them to submit manuscripts to The Professional Geographer or the Annals; 4) invite those in business, industry, and government to interview job prospects at the AAG Annual Meeting or to become AAG corporate sponsors; 5) ask them to serve on editorial boards for journals; and 6) invite them to present seminars as guest lecturers in geography departments.

If the discipline and the AAG are to strengthen their positions in national and international arenas, contribute to understanding, or provide solutions to problems that face our global society, we must also look outside our academic geography departments and build partnerships that advance geography to a broader constituency.

※ ※ ※

Chapter 5

CREATING AND MAINTAINING STRONG AND
HEALTHY DEPARTMENTS

✳ ✳ ✳

Many of the past AAG Presidents have served in administrative positions and know, first hand, what characteristics enhance a department's reputation not only within the university, but also among other geography departments, regionally and nationally. In this chapter, these insights are shared beginning with Risa Palm's column in 1985.

✳ ✳ ✳

RISA PALM
March 1985
WHAT MAKES A "GREAT" DEPARTMENT? Why have some departments prospered while others have failed? Does success rest purely on the basis of the accomplishments of the faculty? I would like to argue that although teaching and research accomplishments are vital, departmental success follows from activity that goes far beyond the sum of individual faculty accomplishments.

When I have participated in meetings of the Council of Colleges of Arts and Sciences (a body of Arts and Sciences deans from pubic universities), sometimes I have heard disturbing gossip about geography departments. Of course, part of the reputation of geography departments comes from the opinions the deans have acquired about the discipline itself—perhaps in their own undergraduate or graduate education. Deans are also influenced by their conversations with one another—and pass on generalized notions about the value of various disciplines in anecdotes and jokes.

We have heard some geographers say that if our scholarship is respectable, we have nothing to fear from university reorganizations or retrenchment. Sometimes these same individuals lament that geographers are misunderstood–that we provide a valuable perspective on intellectual issues and policy problems, but we are frequently ignored; or that in publishing mainly in geographic journals we are "talking to ourselves" and not impressing the wider community of scholars. I think it should be obvious that the value of the scholarship of any discipline, whether published primarily in the journals of that discipline or elsewhere, is not sufficient by itself to merit continued funding in a highly competitive environment. A modest demeanor may have been appropriate for Victorian ladies, but will not ensure survival of departments in current circumstances. Contrary to the homily, if we wait to be discovered, our lights may indeed forever remain under the proverbial bushel-basket.

Because we are doing good work, we cannot assume our activities are being noted by college or university administrations. Administrators are often too busy to keep track of all of their departments and faculty and therefore appreciate being informed on a regular basis of our accomplishments. Until told NOT to do so, we should try to provide administration with as much positive information about the department as possible.

While we all nod our heads at this truism, I have found in my visits to departments that few routinely undertake some very simple and effective strategies to enhance their recognition by higher administration. For example, when was the last time your department invited a Dean to a faculty meeting? When a member of your department received some kind of honor, did you organize some form of public recognition such as a reception, and invite your Dean? When you had a visitor to the department, did you make sure that the Dean and Vice-Chancellor met this individual–by setting up an office appointment? Did you issue a personal invitation to the visitor's talk–something more than a mimeographed announcement of her or his topic? Do you send "good news" from Bob Aangeenbrug in the chair's newsletter to higher administration on a regular basis? Do you try to make sure that you are represented on search committees for new administrators, to ensure that new Deans and Vice-Chancellors will be sympathetic to geography, and that qualified geographers are nominated for such positions?

These are just a few of many possible suggestions. They are simple, but not easy. They demand time and persistence. However, if department members seek the welfare of their departments as well as that of the profession, the benefits of this effort will far exceed the costs.

RUSS MATHER
December 1991

THESE ARE NOT THE BEST of times for higher education in our country. Following a period of expansion of programs and facilities, the current recession has forced most state-supported institutions, and some private colleges, to face the recurring truth that good times do not last forever. A day of reckoning is at hand—when individuals, businesses, universities, and even governments are being forced to retrench. As I traveled the country this fall and talked with faculty and administrators in many regions, it became evident that this is not a problem of just one region or of short duration.

In my twenty-four years as Department Chair at the University of Delaware, there were at least three periods of budgetary stagnation—accompanied by hiring freezes and plans for how we would handle a 5 to 10 percent departmental budget cut. The current economic "climate," however, appears to be much more serious. Consider that salary increases have been eliminated in some schools and budget cutters are terminating academic programs and even restricting "sacred" budgets, such as those for libraries! While geography has had significant success in recent years (e.g., a new Institute for Geographical Sciences was established at George Mason University; a new geography department was founded at the University of St. Thomas in St. Paul, MN; new geography doctoral programs were put forward at the University of Southern California and San Diego State University; the PhD program at the University of South Carolina was expanded; and chairs in geography were endowed at the University of North Carolina-Chapel Hill and San Diego State—all announced in this past year), there is still the possibility that current financial problems in academia may threaten one or more programs in geography.

A recent *Forbes* article emphasized a growing understanding among university administrators that they must begin to consider efficiency, productivity, and the "bottom line" as corporations have long been doing. The article quoted a Harvard economist saying, "The smart thing to do, if you have the courage, is to find your weakest part and eliminate it." Perhaps more than ever before we should take such rhetoric seriously and ask whether any geography departments are at risk. While I hope not, it behooves us to be aware of the problem and to do all we can to see that geography departments are not eliminated. Even if your department is not one of those threatened, you cannot afford to sit back smugly, for elimination of a geography program anywhere may send signals to administrators to contemplate similar action in their universities.

What can we do as individual faculty? First, we can stop thinking in terms of "business as usual." These are trying times and the successful will be those who succeed by trying—to improve their productivity through research, teaching, and service. Successful departments are comprised of productive individuals. Consider what deans and other administrators look for in a strong department and ask how you can contribute to the strengthening of your own department. The possibilities are myriad. For example, individuals can step up their own research efforts, submit more research proposals, increase publication efforts, offer new courses, increase enrollment limits in courses that are usually closed out, or undertake more service activities on university committees. But, a colleague may say, these all mean that we must work harder. That is precisely the point. Geographers must endeavor to become more significant contributors to departments and universities and this takes increased effort, not "business as usual."

Strong geography departments reside in many institutions and they should make an effort to assist their less robust neighbors. Assistance can take the form of exchanges of ideas among faculties, exchanges of individual faculty to present seminars or lectures, joint field trips for students, joint research projects, and even the sharing of certain facilities that might increase opportunities for the less well endowed program. I have touched on just a few of the many possibilities. The time to begin to do these things is now. Let us resolve to do what is necessary to ensure that we come out of this retrenchment period with all our geography programs intact and strengthened.

STEPHEN BIRDSALL
December 1994

PUBLIC COLLEGES AND UNIVERSITIES have not fared well recently in state legislative budget battles. This has affected all disciplines as institution administrators face increasing costs and stable or shrinking budgets. The general prediction for the remainder of the decade is more of the same.

There is a great deal related to this that geographers can and must do during the remainder of the decade: demonstrate your program's quality; be involved in administrative, curricular, and planning decisions; and teach, conduct research, and perform service in ways that are demonstrably unique and valuable.

I am convinced that another exceptionally promising route for geography is university development. In this context, development refers to fund-raising and the connections that are built between university programs and potential supporters of those programs. This route is pursued effectively by a few of our departments, but far too few.

Consider the benefits. The security of endowment income for faculty, students, and programs; financial flexibility; and funds for special initiatives are the most obvious. Other benefits flowing from development efforts are: better relations with your alumni, alumnae, and other supporters; the potential to strengthen or expand your program; an improved understanding of the departmental activities most valued by friends outside your institution; providing students with innovative learning experiences; and, when funds are received, tangible confirmation from outside that your are recognized as an important part of the college or university.

What are the keys to successful development? There is much more than I can cover in this column, but the place to begin is your institution's development office. The most accomplished college and university development officers, while frequently overextended, can be very good and helpful. A good relationship is crucial. You need an advocate who knows your needs and can help you identify your best prospects. When approaching these professionals for assistance, you should come with at least four principles in mind.

First, donations will be made because something or someone in your program is or was very special in the life of the prospective donor or because of an interest developed later in life. The special experience may be a warm remembrance of undergraduate or graduate years, a favorite professor, or a love of geography that grew specifically because of your program and the people who taught in it. Special people seem to have the greatest effect on former students' lives. It is the teaching of individual faculty, teaching in all its forms and perhaps decades after last contact, that is remembered most fondly. Other friends of the department, including corporations and foundations, might focus on your research projects as demonstrations of excellence, or on outreach programs that show the department's unique qualities of imagination and commitment.

Second, potential donors need to be aware of your program's needs, and they need to be asked for support. The goal here is to match donor interest with program need, and this is rarely prepared for quickly. Why should an alumnus care about the department? What are you doing that is special? What would you like to do? What are your greatest needs? What are the potential donor's interests? Whom would he or she like to recognize through a gift? Can the form of this recognition fit with your program's needs? By thinking about how to describe your needs in terms that may be attractive to someone outside the program, you will be better prepared to identify a project idea that will gain support.

Third, donors must be thanked appropriately for their support regardless of the size of their gift. Remember that current donors may also be tomorrow's best prospects. If the gift is in the form of an endowment, the expressions of appreciation must continue without becoming routine. Donations are an investment in your program, its vision and its reality. I cannot imagine a reason to hold back genuine expressions of gratitude for the support represented by program donations. You will need to report at least annually to donors how their support was used and what difference it made. Invite the best donors and prospective donors to campus to interact with faculty and students.

Fourth, when considering development activities for your program, you must accept that you and your colleagues will be in it for the long haul. A great deal of groundwork is necessary. Clear goals must be expressed simply and cogently. Not every goal will be met. Not every potential donor will choose to provide support. Development works best if ways can be found to include alumni and friends in the life and goals of your program, not just on those few occasions when you want something from them. Any relationship works better when there is reciprocity.

All of this takes time, but the benefits for geography in financially-strapped institutions of higher education are significant.

Stephen Birdsall
May 1995

Higher education in the United States, I am convinced, is well into a transition like no other in this country. It has been underway for almost two decades, but recent political events have accelerated the pace of change. How well will geography do?

This past year, I have offered my views of what we need to be doing. At times, I have spoken directly to the issue. In other columns, I have raised the issue only indirectly by discussing related matters. I am aware of having pleased some colleagues, angered others and, no doubt, generated a number of shrugs. Nevertheless, it is a conversation we must have.

Here is what I have been saying.

First, teaching matters. Scholarship takes many forms. All are forms of learning. Some learning is shared almost exclusively with colleagues, but other learning is shared with students or public audiences.

The most effective teachers are those who use their creative scholarship along with an acute sensitivity to modes of student learning. These teachers fashion approaches that inspire, not just inform. The loyalty to the discipline and its representatives engendered through inspirational teaching and masterful public speaking is a priceless contribution that can not be overestimated.

Second, external communication matters. Although growing in number, we remain a small discipline. Our actual and potential contributions are far beyond what has been recognized. We need others, non-geographers, to speak and act knowledgeably on our discipline's behalf because they value what we do. Substantive scholarship of high quality and evident utility, effective teaching that generates a growing student demand and dedicated service both within and outside the institution will accumulate broad support that is invaluable.

Third, a measure of collegiality with geographers matters. We must contend with each other to move our ideas forward. Yet our disputes can be and often are misunderstood by those who know little and care less about geography. When the exchange becomes personal, it serves only the short-term selfish interest of the participants. There are times and ways to disagree with each other that do not harm our collective enterprise. During times filled with administrative challenge, those seeking reasons to cut budgets will often have little patience for programs with faculty engaged in general and obviously unconstructive squabbling.

Fourth, geographic education matters. Better pre-collegiate geographic education improves the quality of students who choose geographic professions, and thus benefits us as a discipline. It also benefits one of our most fundamental goals—to help American society be more geographically literate.

Geographers not yet involved in the effort to improve K–12 geographic education but who acknowledge the need for this effort can begin to contribute in many ways. Talk to your state's Geographic Alliance representative. If you don't know who it is, ask Osa Brand in the AAG office. Visit with and listen to one of the thousands of pre-collegiate geography teachers in your area talk about what he or she needs and what he or she finds exciting about our field. Look carefully at "Geography for Life," the national geography standards, and consider how they might help your own undergraduate teaching. Talk with School of Education faculty to learn how decisions are made locally about what subjects are taught in the public schools. Act.

Fifth, and most importantly, quality matters. One of the most challenging aspects of the change underway in higher education is to develop a clear-eyed view of the quality of our own efforts. Most of us believe that critical evaluation of one's work by qualified peers is the best way to ensure quality. Refereed manuscripts and grant proposal review panels are two established examples of critical peer evaluation.

I agree that this system is the best yet developed for scholarship, but it has received more than usual criticism recently. Something, it seems, is missing.

Some criticism speaks to apparent ethical lapses that occur in the review process from time to time. Even more problematic is the suggestion that when assessing quality, peer review does not pay sufficient attention to matters of social value. While this latter concern begs for much longer treatment than one of these columns, at the very least it suggests that we not define quality solely according to the isolated standards of "pure" scholarship. Much of what we do has to matter to others outside the academy as well as to those of us within. This is partly a matter of teaching and external communication, but we also should remain mindful of the need to connect our teaching and scholarship to the world of others.

Geography as a discipline is well placed to ride the transition underway. We will not emerge unscathed, but if collectively we attend to these five issues and take advantage of our field's interdisciplinary, international, and pragmatic traditions, geography will come through the period ahead stronger than ever.

Judy Olson: *Evil (?) Administrators*
September 1995
HAVING RECENTLY FINISHED a five-year term as chair of the Michigan State geography department (an appointment my colleagues and I often jokingly referred to as a five-year sentence to the chair), I have reflected on relationships between faculty and administrators and the attitudes they harbor toward each other. I recently queried a colleague about programs where graduate students were acquainted with the mechanics and principles of university administration. The subject line of his e-mail response was "Evil Administrators," suggesting he had divined my concern: certain prevailing attitudes of faculty toward administrators and the need to provide academically-bound graduate students with insights that will serve them better.

Faculty sow attitudinal seeds that bloom perennially, even decades later when their students are asked to accept administrative appointments. I cannot number the times I or a colleague have said "When I was in graduate school . . ." followed by an adamant position on the issue at hand. Such words betray where their speakers come from—literally as well as figuratively—and they explain much current thinking. I don't recall ever being told in graduate school that every faculty member should take a turn in the department chair. Yet finding myself in a department with changing leadership 19 years later, it seemed to be the appropriate thing for me to do.

Once in the chair, I came to realize how different my attitudes and preparation for the job might have been. I also became sensitive to how counterproductive prevailing faculty attitudes toward administrators could be. Casual comments and conversations (inside and outside my department; at my home institution and others) are revealing: faculty perform the noble roles of teaching and research in universities; administrators are not bad at best, are sometimes useful, and only occasionally worthy of plaudits. For the most part, administration is a necessary evil; one might dabble, but should never get caught up in it. Administrators are them and we are us. Someone interested in administration is suspect.

I exaggerate. If such negative attitudes dominated my department I would not have survived. I shall be grateful forever for my colleagues' support and understanding. Yet the negative attitudes that exist serve us badly as faculty in at least three ways: 1) they make the transition from faculty to administration difficult for individuals; 2) they demean those who hold them and the institutions they serve; and 3) viewing administrators as adversaries hampers institutional teamwork.

Despite the prevalence of negative attitudes, however, some colleagues in geography have made the transition temporarily or permanently, and they seem none the worse for it. In fact, Ron Abler characterizes chairing a department as a "transforming" experience that works a positive metamorphosis on most who undertake the task.

I asked some geography colleagues who like administration about their thoughts.

*Debbie Elliott-Fisk, Director of the University of California's Natural Reserve System, enjoys administration because of its potential for making a difference and effecting change. How did she acquire a positive attitude toward administration? She replied with no hesitation that in her early days as a faculty member, she was exposed to competent, visionary administrators committed to their institutions and to students. Their attitudes and enthusiasm were infectious.

*Jack Vitek, currently Interim Associate Vice President for Academic Planning at Oklahoma State University, has served successfully in administrative posts where he focuses on students. His long experience as an advisor prepared him for his role in administration. As a faculty member, he could influence only his own students; as an administrator, he can improve the lot of entire student bodies. He admits his initial entry into administration was dumb luck. More recently, he saw a clear opportunity in a position that would enable him to move his institution in the direction he was convinced it should go. He now feels comfortable with faculty or administrative roles, or combinations thereof.

* Steve White presided over the Kansas State University Department of Geography as it won its new PhD program. He admits that his initial encounter with chairing was unplanned, a hurried decision at a tender age that fortunately turned out well, as it lasted eight years. His recent reentry into department leadership has been comfortable and enjoyable, even though the six-year interlude on the faculty was also rewarding. With

support from his colleagues, Steve is able to include teaching and research on his agenda as well as administration.

These positive stories could be multiplied tenfold or more. The AAG Academic Leadership Round Table consists of 90 geographers who are deans through university presidents. Geographers who have chaired departments during their careers must number in the hundreds. Most did well by themselves and their units. Most probably began with negative attitudes toward administration and changed them.

Why burden the next generation with the prevailing skepticism about academic administration? Graduate faculty would do university-bound students a major service by developing them into more effective academic team players. Many students will assume administrative roles at some times during their careers. Why not acquaint students with the accomplishments of dedicated, effective, and visionary academic leaders such as those who inspired Debbie Elliott-Fisk? I have no wish to transform geography into an administrative training specialty (though geographers may possess traits that account for their presence in higher administration). But future faculty would be more effective if they knew how to work with rather than against academic administrators.

PAT GOBER: *Grassroots Leadership*
July 1997

LET ME SAY at the onset of my first presidential column how honored and humbled I am at the prospect of addressing the AAG on matters that are significant to me and hopefully to you as well. Enough people have told me that they are awaiting my columns to make me realize that our members read, mull over, and react to the ruminations of their leadership. I plan to use the opportunity of the next twelve monthly columns to share my ideas about the future of our discipline and the role that each of us can play in shaping a robust and bright future. This month I will discuss the importance of leadership, especially at the grassroots departmental level.

Since my earliest years in the Association, I have heard the lamentable stories of geography department closures. I cut my teeth as a departmental chair during the panic years of the early and mid-1980s when Michigan, Columbia, Pittsburgh, Chicago, and Northwestern closed their Departments of Geography. In spite of the emergence of new programs and a healthy overall outlook for the discipline, there are several geography departments today in jeopardy of elimination, merger, and restructuring. Their members understandably look to the Association for support in trying to avert these administrative actions.

Attempts to buoy sinking geography departments are, in my opinion, doomed to failure. Short-term fixes are too little, too late. The real answer lies in cultivating long-term leadership for building and maintaining healthy geography programs. The 380 baccalaureate, 170 masters, and 75 PhD programs in North America are the bone marrow of the discipline. They must be continually nourished and renewed so that they can produce a healthy new crop of geographers.

Maintaining strong geography programs in the long term requires real leadership at the departmental level. It takes persons who can articulate what geography does and why it is important—persons with the judgement, courage, imagination, and creativity to see how geography programs can take advantage of trends in higher education rather than being swamped by them.

Departmental chairs are in pivotal leadership positions. They mediate between what is important to the college or university and what the department wants to accomplish. One of the surest ways to stymie a department's program is to choose a weak chair who cannot articulate what the department contributes to its college or university, who cannot be bothered with the "vision thing," and who is simply too busy for the mundane, day-to-day business of making the unit work.

Rarely however, do departments have the luxury of a continuous string of strong and effective chairs. Sometimes faculty members, even junior ones, need to step up and take the lead. Effective departmental leadership comes from a variety of sources, including the chairs of personnel, graduate, and curriculum committees, an undergraduate advisor, a faculty senator, or a junior faculty member who sees the need for curriculum reform and presents the faculty with a viable proposal. In other words, we all need to take responsibility for the health of our geography programs and we need to do it on a continuous basis.

Although these are largely matters of personal and departmental responsibility, there are several things the Association can do to promote grassroots leadership.

1. Regularize leadership and network-building activities among new faculty members in geography programs building upon the success of the Phoenix and Miami Groups. We need to ensure that junior

colleagues take ownership of their departments and their discipline rather than isolate themselves in the single-minded quest for tenure.

2. Use the Annual Meeting Chairs Luncheon as a leadership-building activity. We might provide the usual opportunity for networking but expand the formal program to include planning, management, and leadership activities.

3. Sponsor summer leadership workshops to identify trends in higher education, geographic research and instruction, curriculum reform, personnel management, and problem-solving strategies. Cultivate a critical mass of visionary geographers who may or may not hold the position of department chair.

These are heady but risky times for geography. Our collegiate departments are experiencing unprecedented growth in enrollment, geography has been designated for special emphasis in America's schools, GIS has opened new employment niches for our graduates, and the discipline is in increased demand as an approach to science and environmental issues. At the same time, the pace of technological and institutional change makes it imperative that departments not stand still, rest on their laurels, and congratulate themselves on rediscovery of geography. We must find ways to build upon our past success, to add programs to our critical mass, and to bolster the programs that we already have. The AAG can build an infrastructure that facilitates leadership, but the future success of our geography programs involves leadership at the local level.

REGINALD G. GOLLEDGE:
Some Basic Survival Skills in the Academic Jungle
November 1999

Geography's image is steadily improving. But there are still threats to the continued existence of some departments. Counteracting these threats is no simple matter. In this column I continue the tradition of AAG Presidents offering suggestions for mitigating those threats.

Academic Quality

There is no substitute for the academic quality of your faculty. But even departments with quality faculty have been closed or downgraded to program status. Two strategies appear to dominate faculty hiring—"fill the gap" or "take the best applicant regardless of interest area." The first indicates hiring is dominated by curriculum interests. Sometimes this is necessary, sometimes it represents a new growth interest, but sometimes it reflects inflexibility and a reluctance to change. The latter can be a dangerous principle in times of significant social, economic, and academic flux. The second alternative shows a clear commitment to academic quality, a willingness to explore new curricula, and an understanding that change may be necessary to reflect current and future trends. This often earns an evaluative plus both from the administration and students. Suggestion: hire the best and, if needed, change your program; don't get stuck with meeting out-of-date needs.

Teaching Quality

Not too long ago, teaching was sometimes regarded as the "necessary evil" of an academic position. Not today! For more than a decade, teaching quality has been increasing in importance in academic evaluations. Teaching quality includes both course characteristics (interesting, relevant, and useful content; current and relevant texts and readings) as well as instructor characteristics (good presentation skills, use of interesting and appropriate teaching technologies). No longer is it desirable to penalize non-researchers with higher teaching loads unless you are confident of their teaching ability. And for those who prefer to teach rather than do systematic research, encourage them to write about their methods of preparation and presentation. Research in geographic education has escalated in importance, particularly since the "Standards" and "Geography For Life" projects have provided such excellent bases on which to build advanced work. Neglect of teaching quality and consequent poor student evaluations can imperil a graduated department almost as much as lack of research. Lack of teaching quality in a four-year college with less emphasis on research can be the kiss of death.

Students

Department budgetary allocations usually are driven by course enrollment and number of majors. While some nationally respected departments continue to survive with good graduate programs and small undergraduate majors, they often do this by offering (required) General Education (GE) courses at the undergraduate level. This can be dangerous—a campus move to change GE requirements can quickly erode a department's student FTE base. As many chairs recognize, relatively few students enter college as declared geography majors. More usually, exposure to geography via an elective or a GE course turns them on to the discipline. Departments should have an active rather than passive program to encourage this change. Active programs might include interesting field trips, a UG geography club, honors programs, meaningful coursework that contributes to employability, and a good internship and placement program. The latter often can be enhanced by a good alumni program.

Campus Politics

Knowing what is "happening" on campus is important. Knowing before other people know is an essential part of departmental offense and defense strategies. Faculty strategically placed on campus-wide administrative committees, and (in some cases) academic senate committees, can provide invaluable advance information on new initiatives, supplementary (competitive) campus funding opportunities, and other ways to access extra resources. Defensively, early access to information can help a department prepare materials in advance of rumored or actual attacks. A good chair should know all the powerful committees and try to ensure departmental representation on all of them. Examples may be Research Committees, Academic Planning, General Education, Academic Personnel (Hiring) and Evaluation Committees, Library Committees, and Educational Policy Committees. Don't neglect this—it might be of crucial importance.

Research Funding

Every college administration loves extramural funding. The overhead so generated gives an added flexibility that is otherwise unavailable. Some academics have a philosophical objection to seeking extramural funding, arguing that academic research should be unsullied by things such as constant supervision, restrictions on publication of results, and pressure to bias findings in favor of the funding agency. Agreed. But this is far from the typical scenario for research funding. Note there is a difference between "contracts" and "grants." The former often can be obtained by acting as a consultant, or accepting funding from agencies (such as the Department of Defense) which have clearly stipulated reporting, oversight, and disclosure requirements. Grants are usually obtained from philanthropic foundations or from governmental funding agencies (e.g., NSF, DOT, NIH). These usually allow individuals to submit a proposal to do research on a problem in which the individual is interested, not one dictated by the agency. Funding depends on how your peers evaluate your proposal. Often competition is open with no thematic restrictions (e.g., NSF Geography and Regional Science Program); other times you must find an agency whose Request For Proposal (RFP) best matches your interest (e.g., K–6 teacher training; equipping departmental laboratories). Overall, extramural funding activities are a plus for a department (as every poorly funded graduate student can appreciate). Let every faculty member be driven by their own conscience in procuring grants and/or contracts, but use campus and AAG resources to help establish an information base, an environment, and a presence for extramural funding in your department. Ignore the naysayers who may never have prepared or submitted a proposal. If your ideas are judged worthwhile by your peers, you'll get funded—maybe not on first try (I didn't), but you can learn the art and science of proposal writing.

Alumni

Never disregard or underestimate the value of your alumni. After all, you trained them and sent them into the job market. Keep in contact. Establish a newsletter for alumni to keep in touch with department events and with each other. Your students will benefit from knowing what type of future employment is open to them. You may benefit from unexpected gifts and bequests. And alumni are often a powerful ally in a fight for existence.

Planning

Plan for your future. Even if your administration doesn't require it, prepare a 5-year plan, setting out goals and objectives. Plan for changes in curriculum and for participation in new directions taken by the discipline. Plan for new faculty to help pursue your objectives. Examine your efforts in pursuing equal opportunity and minority representation goals. Revise your plan every two years, whether you are asked to or not. Show your administration that you are forward looking and aggressive in pursuing your goals.

Marketing

Never be ashamed of (or reluctant to admit to) being a geographer. Work on improving geography's image to make it compatible with what the discipline really is today. Engage in community outreach. Get department members out to local schools during Geography Awareness Week. Invite local business leaders to a "show and tell" each year to demonstrate the nature of your product—well-trained, thoughtful, and independent workers. Develop intern programs. Advise local businesses that they can advertise their jobs in the *AAG Newsletter* or online. Make sure your program is producing students with skills and knowledge that is wanted by employers. We can't all just produce academics.

There are many other dimensions that could be stressed for helping your department to help itself. You don't have to be in danger of closure to benefit from these suggestions. Be proactive, not re-active. Overall, don't be complacent and think that because you've survived in the past and that the "worst is over" in terms of potential department closures, that you can drift along doing just as you've done before. Do that and you'll find that aggressive activity by other departments will erode your position. Other departments will start GIS labs; other departments will emphasize computer graphics or cartography; other departments will enlarge the "spatial" aspects of their domain, leaving geography with a restricted core and restricted attractiveness. Don't be complacent! Remember all the highly ranked departments that have been terminated in recent decades.

M. DUANE NELLIS: *The Successful Geography Department*
August 2002

Over the past fifteen years I have had the opportunity to serve as an external evaluator for several geography programs around the country. These programs have ranged from quite small (three faculty) to some of our largest (over twenty faculty), and from undergraduate only to those offering degrees through the PhD Some have been joint programs and others have been standalone programs. Many have focused on "the right things," while others have struggled to define themselves and their role at their respective universities. I also have the perspective as dean of a college of arts and sciences. In that role, I examine programs beyond geography and evaluate what makes certain units outstanding and what places others more at risk. I thought it might be useful to share some of my thoughts and experiences with you in the hope that such a perspective might be of value.

One key characteristic of a successful geography program is a clear vision of what the department is and what it wants to become. Such departments have a common agenda (or at least a non-conflicting agenda), with everyone within that department working in an agreed–upon direction. Such a focus or vision has become even more important in these times of budgetary constraint in higher education. Here at West Virginia University, the geography program has developed its graduate programs around the fields of geographic information science, development and planning, and resource and environmental geography. When I was head of geography at Kansas State University, we were able to build a new PhD program plan around the area of rural geography. Some geography programs I have reviewed simply want to be strong undergraduate programs with high quality teaching that broadly prepare students for careers in applied geography or for geography graduate programs at other institutions. As faculty debate what their program's niche should be, they should consider the type of institution they are a part of, their institution's mission, and the complementary and supportive programs outside geography and across the campus. The geography department at Oklahoma State University has certainly been sensitive to doing what is right in this regard, and was rewarded for doing so when the Oklahoma State Higher Education Governing Body recently approved their new PhD program.

Having adequate resources is another characteristic of a successful program (budget, facilities, people, and equipment). One often hears, "If the dean and provost would only give us such and such, then we could do x." But, of course, the reality is that budgets at most universities and colleges nationwide are getting tighter, and successful departments are expected to be (and often are) more entrepreneurial. Strong efforts to build extramural support (for both research and instruction) and to strengthen development can make a big difference in helping geography programs move from marginal levels of resource availability to more significant levels. Other support resources, such as having a great administrative assistant and adequate technical staff support, are also common successful geography programs.

A student-friendly/student-centered environment, and a strong commitment from faculty toward their students, is another common theme in successful programs. The most successful geography programs often have an active Gamma Theta Upsilon Chapter or some type of geography club that engages students in a variety of learning experiences both inside and outside the classroom. These programs and activities create a sense of community among the students and faculty, and provide a link to what is happening in the profession (e.g., through a visiting lectureship series—even a coffee hour—or simply through informal discussions on various topics, including career opportunities in geography). Such an environment generally includes quality advising of both undergraduates and graduate students, and a curriculum that is grounded in the fundamentals of our discipline and at the same time reflects the exciting changes in our approaches to solving geographical questions. Student-centered departments also offer challenging and quality courses to non-majors as well as majors, and have worked to ensure that certain geography courses are part of the general university requirements. Placing students in quality graduate programs (and PhD graduates in high quality university, government, or private jobs), geography-related jobs, or in internship opportunities also is common in successful programs. And geography programs that have strong assessment instruments to measure student and graduate student success are positioning themselves well for positive feedback.

An effective leader, someone who is a politically astute department chair or head, is also an essential and common trait in successful programs. An enthusiastic leader who works hard to ensure faculty are rewarded for their contributions makes a big difference in creating an atmosphere for success. A geography administrator who is fair and open-minded, provides constructive feedback, makes decisions, and works with faculty to create a sense of direction is also critical. Of course, having a chair or head who is well-connected and respected on campus helps tremendously as well.

A sense of collegiality is another characteristic of successful geography programs. This doesn't mean that the faculty needs to agree on the multitude of issues that departments grapple with, but that they can agree to disagree, they are not threatened by having personal views challenged, and they maintain and foster a sense of respect for one another. In threatened geography departments, it is often faculty in-fighting that leads to such programs becoming targets for downsizing or elimination. Efforts to settle scores with other faculty members by encouraging student unrest or carrying conflict among faculty to the entire college or campus community is both shortsighted and dangerous.

Successful geography programs often have an environment that is conductive to a "can do" rather than "we have already tried that" spirit, which can go a long way in creating an atmosphere that is supportive of (and rewards) innovation and risk taking. This includes sharing burdens and responsibilities, and holding regular faculty meetings to discuss the broad range of issues and opportunities associated with an active geography program. Successful programs often have faculty meetings that appreciate diversity and different points of view, but that also nurture a sense of moving in a focused direction. Certainly a part of that shared focus is the belief that if the department as a whole succeeds, all faculty members benefit. In a number of cases, for example, geography programs have benefited from the additional investment and interest in geographic information science. Such investments have led to new commitments from deans and extramural funding agencies, provided greater recognition of the department on and off campus, and created excitement among (and new opportunities for) students. Yet, occasionally in my reviews some faculty have expressed reservations about this new emphasis without being willing to examine the benefits such new developments have brought to their program and to the discipline.

Faculty active in quality scholarship with a strong research base, who are active (and encourage their students to be active) professionally, have academic colleagues across campus and beyond, who publish in good quality journals, who are committed to quality and innovative teaching, and who are able to use their active research programs to feed exciting teaching, are also common variables in the most successful geography programs. Having faculty who are also willing to reach out to the K–12 community (e.g., through the Alliance network) and build partnerships that are more service-oriented can be another plus.

Other characteristics of a successful geography program include the department's position as an integral part of the campus community, opportunities for faculty development and mentoring, strong links with alumni (including regular contact through a department newsletter and some type of annual alumni event), good connections to the local community (and, where appropriate, with key state and federal agencies), a reasonable and effective department committee structure, and shared responsibility of service obligations within the department.

There is no simple model for a successful department of geography, but I hope some of these observations will be of value. As universities and colleges face tighter budgets, higher levels of accountability, and stricter measures of productivity, it will be important for geography programs to be sensitive to ensuring their long-term positions as a core department within the range of disciplines that are central to a university's success.

※ ※ ※

Chapter 6

DIVERSITY ISSUES

❋ ❋ ❋

Diversity issues—intellectual and human resources—have been and no doubt will continue to be a major concern of the AAG and its elected Presidents. The improved status of women within the profession, the continued lack of under-represented groups in the discipline, and the need for a catholic understanding of our intellectual richness provide the fodder for this group of presidential musings.

❋ ❋ ❋

RISA PALM
October 1984

WITH A WOMAN CANDIDATE for national vice president, and a woman justice of the Supreme Court, we are tempted to agree that indeed women "have come a long way." Because I am personally acquainted with a surprisingly large number of women geographers who seemed to be on a path of professional success only to find themselves denied re-appointment or tenure and promotion during the affirmative action days of the 1970s, I was doubtful that this statement actually applied to professional geography in the United States. I decided to try to gather a few statistics either to back up my impressions or disabuse me of my errors.

I decided to compare the current status of women with that described ten years ago by then-AAG president Wilbur Zelinsky in "The Strange Case of the Missing Female Geographer" (*Professional Geographer*, 1973, pp. 101–105 and 151–165). I limited myself to what Zelinsky called "the more rarified levels of academic geography," where he described the outlook as "particularly gloomy." I tabulated the numbers of female faculty at various ranks in US departments offering advanced degrees listed in the 1983-1984 *Guide to Graduate Departments of Geography in the United States and Canada.*

The past ten years of affirmative action have indeed seen increases in the number of women geographers in academic positions in institutions offering advanced degrees. Zelinsky found only four full professors and thirteen associate professors among the 1,729 US and Canadian geographers listed in the 1971-72 *Guide.* A decade later there were seventeen female full professors and twenty-three associate professors among the 1,981 geographers for whom ranks were reported. In the US departments where ranks were reported, almost sixteen percent (47) of all the assistant professors were women, four percent (19) of the associate professors and almost 2.5 percent (16) of the full professors were women. The US PhD-granting departments had a higher percentage of women assistant professors (23.7%) than Canada, a similar percentage of associate professors (5.4%), but a low percentage of full professors (1.8% or six of 337).

Zelinsky reported that informal conversations had given him the impression that greater numbers of women were entering graduate school in the early 1970s. The list of PhD dissertations presented in the 1982-83 Guide shows that 24 of 84 (28.6%) PhDs reported that year were done by women. Although women are entering our graduate programs, and successfully completing the PhD, a smaller proportion is employed as assistant professors, and fewer still attain promotion and tenure.

There have been few increases in what Zelinksy called "formal recognition from the geographic professions" for the efforts of women. In 1973 only four women had received honors from AAG. By 1984 that number had increased to six. In 1973 only three women had been elected to the AAG Council; by 1984 eleven women had been elected, and women had served as executive director. In 1973 only one woman had ever been president of the AAG (Ellen Churchill Semple in 1921). The number has now doubled!

What do these numbers tell us? Is geography a field where women can progress significantly? To be fair, the results are mixed: the number of women at lower academic ranks and serving as AAG councilors has increased. On the other hand, a profession with only six female full professors in the 53 PhD-granting departments in the United States can hardly be proud of its affirmative action record. In a subsequent column I will explore some of the issues behind these statistics.

Risa Palm
February 1985

In the 1 October, 1984 AAG Newsletter, I presented some statistics showing slight increases in the number of women geographers in academic positions in institutions offering advanced degrees. These numbers reflect some increases at lower ranks but virtually no change at the highest ranks—in 1983-84 there were still only 6 women vs. 331 men holding the rank of full professor at PhD-granting geography departments. I promised to explore some of the issues behind theses statistics in a later column.

Since tabulating those figures, I have had the opportunity to visit many regional meetings and to learn more about attitudes towards female colleagues held by both men and women in various geography departments. I believe that progress is being made, and that when an interested person re-tabulates these figures in 1995, there will be larger percentages of women at all ranks. However, there are still problems to be overcome— and measures that can be taken by departments and university administrations to increase the diversity of its faculty—including the recruitment and retention of women.

What are some of the current problems? I would argue that neither overt discrimination nor lack of interest in geography by women is among our prime problems. At present, discrimination, although possibly present, is far less of a problem than the blatant attitudes of the early members of the AAG described by Allen Bushong (*The Professional Geographer* Vol. 26, 1974, p. 437) when members actively sought to exclude women from evening sessions (known as "smokers"), and went so far as to avoid particular topic areas for discussion in order to decrease the likelihood of attendance by women. Second, although geography has been identified as a predominantly male area of intended study by college-bound seniors (*Women Today*, Vol. 14, October 26, 1984), there is evidence that more than one quarter of all PhD dissertations granted are to women and that women make up large percentages—sometimes more than 50%—of the graduate student populations in some of the larger graduate departments.

Instead, the problems facing women in academic geography are similar to those facing women in other academic fields in which women are not already well represented. In geography departments with only one woman, the woman may suffer isolation—she may find it difficult to participate effectively in a wide range of collegial behavior. She may not know how to cope with numerous requests to serve on recruitment and other committees requiring women representatives. She may face unfounded suspicions by spouses of departmental members, or greater scrutiny given to her living arrangements than that given to her male counterparts.

Even more generally, all academic women—whether they are isolated in otherwise all-male departments or not—face conflicts between childrearing responsibilities and their academic careers. Women must make decisions concerning the allocation of their time and energy particularly during the critical first seven years as assistant professor—a time when peak productivity in research is expected and yet also a time when the responsibilities of caring for a young family are most likely. In addition, since more academic women than men marry at or above their own level of education, they are more likely than academic men to have spouses with high levels of education and high career expectations for themselves. This results in the familiar dilemma of the dual-career household and the possibility that a greater proportion of women than men move to a less than ideal location for their own careers, simply in order to accommodate the needs of both spouses. Of course, many such problems and conflicts are also shared by men: the point is that they are more frequently encountered by women than by men, and, therefore, are more likely to have an adverse effect on the careers of women.

What can departments and universities do to increase their success in recruiting and retaining women? University administrators and department chairs must gain an awareness of the individual vs. societal nature of problems encountered by women in the academic setting. Administrators should consider changing regulations that place women in almost impossible time conflicts in their early years at an institution—asking less service commitment from untenured women and modifying the seven-year lock-step advancement to promotion and tenure. Departments and administrations may be more open to job-sharing. Department members may make more efforts to include women in informal gatherings. Departments may also wish to help women students and faculty to keep in touch with other women geographers by supporting programs of departmental visits of women faculty from nearby institutions.

The problems of increasing the numbers and success rates of women in academic geography are large—they go far beyond any situation unique to geography. However, they are amenable to solutions and there are potentially great rewards for efforts to increase the number of women in academic geography. By including more women in our field, we will enjoy a significant infusion of new talent as well as the prospect of new perspectives on classic geographic problems.

SUSAN HANSON: *Musings on Multiculturalism*
November 1990
PLURALISM, MULTICULTURALISM, DIVERSITY—all are current watchwords in business, government, and education. Colleges and universities are making "educating for multiculturalism" central to their mission; with a strong tradition in this arena, geographers have a great deal to contribute.

Some time ago the melting pot was abandoned as an apt analogy for US society in favor of a lumpier concoction described variously as tossed salad, chunky stew, or vegetable soup. What these culinary metaphors were meant to conjure up was a society whose very richness lay not in a smooth, homogenized puree but in a diversity of textures and compositions, each of whose special contribution to the mélange was to be celebrated. No matter if the flavors sometimes clashed; the value was in each ingredient's uniqueness.

In the days when my kids enthusiastically (and I even more enthusiastically) watched *The Electric Company*, PBS also ran a children's program designed to prepare its young viewers for living with tolerance in a multicultural society; each episode of *Vegetable Soup* highlighted the culture of a particular ethnic group in America. Since then, the variety of "vegetables" in the American "soup" has grown, not diminished, and, as differences have become not only appreciated but also feared and shunned, multiculturalism has moved over more deliberately to the center of American life.

In higher education, interest in multiculturalism has manifested itself in a plethora of "_____studies" programs: Afro-American studies, Chicano Studies, Women's Studies, Men's Studies, for example. Focusing largely on differences, these programs have enriched the college curriculum by forcing us to recognize and explore diversity. But a confrontation with difference should, by necessity, entail as well confrontation with the other end of the continuum, similarity. With its focus on space, place, and location and with its interest in difference as well as with the ties that bind, geography is well suited to educating Americans for life in a distinctly unpureed society.

On the one hand, understanding and appreciating difference is what geography is all about. In the memoir of her childhood in New South Wales, *The Road from Coorain*, Jill Ker Conway underlines geography's emphasis on place-to-place differences in the following incident. As a university student in Sydney, she goes to a play in which she sees for the first time the experience of "real Australian types" being treated as if they were universal, and she reflects, "seeing the world that way on stage helped to undo a lifetime of lessons in geography." Place, space, location can indeed be central to forming separate identities, to fostering a sense of community within a group, to dividing one group from another.

On the other hand, geographers also deal with similarities, with universals, with the connective tissues that link together diverse peoples and places. Everyone is caught in the web of space and time, boundaries are fluid and permeable as well as divisive, and space, place, and location can unite as well as divide diverse groups. Geography speaks to differences but also to commonalities and to ways of bridging difference while sustaining it.

As Bronowski so delightfully demonstrated in *The Abacus and the Rose*, his little morality play on the follies of juxtaposing science versus the humanities, all who attempt to understand the world wrestle with both the unique and the universal—and often end up finding the universal in the unique. If multicultural education emphasizes difference, separateness, distinctiveness, and uniqueness without also stressing commonalities, bridges, and universals, the potential richness of life in a truly multicultural society will be lost. Our discipline's long history of attention to both what divides and what unites people and places makes geography an ideal vehicle for multicultural education.

JUDY OLSON: *Feedback on Affirmative Action vs. Privacy*
February 1996

I AM PLEASED THAT SEVERAL AAG MEMBERS responded to the December 1995 column in which I posed questions concerning affirmative action and privacy. In short, the AAG membership form now has a checkbox for minorities who choose to allow AAG to release their names for recruiting purposes. Council has received a proposal for a similar check box for women; gender has heretofore not been confidential information. We will make use of the information received from respondents to inform decisions about changes to the membership form.

The general consensus, not surprisingly, was that members should be able to control whether their names go on a categorical list. The matter, however, is a part of a much broader set of issues associated with affirmative action and privacy, and those issues were reflected in the various responses. In this column I will address some of these broader issues.

The very question of whether there should be affirmative action is one matter that arose in the responses. A wide range of opinions was reflected. Some characteristics of affirmative action might best be clarified, as some see it as a quota system or as downright favoring certain groups. In my own experience as an administrator and hiring committee member, affirmative action has demanded that all due effort be made to identify qualified people in various groups, but never have I been under pressure to hire in a certain category for its own sake. The only time that "preference" has been given is with equally-qualified candidates. Otherwise, the candidate on top, majority or minority, male or female, was the candidate to receive the offer.

Can zealous pursuit of affirmative action goals sometimes result in hiring an unqualified person? I would be surprised if it never happened, but I have always found it interesting to observe how attitudes (about lack of qualified candidates in non-majority categories) change when an effort is made to find them. On the whole, women and minority hires do not seem to have deteriorated the discipline.

Ideally, of course, we should concentrate on qualifications and not categories. Perhaps we will reach the stage when we are truly color blind and gender blind (though arguments could be made for more positive color and gender vision); meantime, we still need to be sure that we are giving opportunities to all. The discipline needs women and minorities not just in fairness to those individuals. The discipline is the richer if more people are allowed to make their fair contributions. I recommend the thoughtful piece by Roger Wilkins (*The Nation*, March 27, 1995, pp. 409-416) entitled "The Case for Affirmative Action: Racism Has Its Privileges" for a more extended discussion—thanks to Joe Wood for calling it to my attention.

All of this idealism, of course, must be less than comforting for any majority reader who has felt unfairly passed by for a job that has gone to a minority or female. Anyone who has ever felt the sting of discrimination is unlikely to wish the feeling on anyone else. We have to keep trying to do things for all, which is all that affirmative action was ever intended to do.

Another issue brought out by the questions concerning the membership form is the matter of self-identification. Let me first make it clear that, above all, I respect the desire for and the right to confidentiality of minority and gender status. On the other hand, I see a clear case to be made for encouraging self-identification, not just on the forms but in other contexts as well. I mentioned in the December column that I was grateful for women who went before me and did not hide their gender. That does not in any way detract from my own professional efforts; it just recognizes a piece of truth: it would have been much harder without predecessors. Being able to see what contemporary colleagues are accomplishing, regardless of (known) racial/ethnic or gender status, reinforces our confidence in one another and dispels some of our tendencies for unfair discrimination. Many of us are pleased, too, that our accomplishments might open the door a little wider for others.

I believe we also send some unintended messages when we hide our identity. I encourage students with gender-neutral or unfamiliar foreign names to indicate gender clearly in correspondence, especially if it is job-related. Even though I would be against any requirement to so identify, I am familiar enough with hiring processes to speculate that the comfort level of those with a vote can affect the outcome. For whatever reason, knowing gender is important to our comfort and to our being able to think about a person's other qualities. I suspect, too, that a candidate's discomfort with his or her own identity, racial/ethnic or gender is sensed by others. If the candidate is uncomfortable with it, perhaps others are as well. Different people do have different strategies for dealing with their own identities, however, and I do recognize that what works for one person does not necessarily work for someone else.

Yet another issue arising from the membership form discussion is how AAG can promote networking between hiring entities and those in the profession likely to know of good candidates. Senior faculty who are not themselves on the job market but who are likely to know qualified candidates will not be accessible through the lists resulting from the membership form unless they check a box to receive job notices on behalf of others. Likewise, departments seeking minority and women graduate students will not be helped by the current and proposed check box unless "recruiting" is interpreted very broadly. It may be useful if we consider other ways in which AAG can facilitate appropriate networking for these purposes.

REGINALD G. GOLLEDGE: *Using Our Human Resources*
August 1999
RECENTLY, WHILE RECOVERING from a medically enforced week at home, I had an unexpected opportunity to take a trip down Memory Lane. Like other AAG Presidents, I was diligently searching for a suitable topic for a presidential column and had decided to write about a lack I had observed of late, namely, the paucity of ways to tap underutilized resources of the AAG. Ron Abler has been urging the profession for some time to do more for our non-academic membership—particularly those members employed in government and business and corporate members. This is a sentiment with which I strongly agree, and past president Graf's invitation to Jack Dangermond of ESRI to give a plenary session in Honolulu was a significant move in this direction. The new AAG Mission Statement emphasizes this need and further steps will be taken to address this objective in this and future years.

There is a substantial section of the Association that represents a resource that we have developed far less than we should. Approximately one in every seven members holds a position outside of the college/university arena. These thousand-odd members include government employees, corporate employees, private business owners and operators, and other groups.

There is no doubt that until recently the bulk of the efforts of the AAG has been directed toward providing services for its academic membership. As part of this service, we run national and regional conferences, publish journals, monographs and technical reports, provide job search information, identify some research funding deadlines, publish the dates of important geography meetings and similar information on related disciplines, and so on. In general, the efforts of the Association are well appreciated. But in this column, I would also like to explore what more can be done to incorporate the non-academic members into our mainstream activities.

Last year, President Will Graf contacted a number of people who had failed to renew their membership in the Association to inquire about their reasons for this action. Many of those who did not renew were non-academics. A major reason given was that the AAG no longer provides services of interest, or no longer serves their needs. This, I believe, is a major challenge that should be addressed now so that in our future, we attract and hold on an increasing number of non-academic members. A question we have to address is, "what services can we provide in order to do this?" The AAG already makes available job information on its website; for 1999 (January 1-June 24) the AAG website had: Total Number of Requests 307,014; Total Number of Visits 78,403; Total Number of Visitors 34,816. The reorganization of the discipline's two major journals after extensive discipline-wide interaction over the last eighteen months has brought forth suggestions as to how the *Annals, The Professional Geographer* and the *AAG Newsletter* can better serve the entirety of our geographic community. It will be a challenge to the future editors and editorial boards to pursue the revised missions of these publications in such a way as to make them attractive not only to the academic members of the AAG, but by providing extended opportunity for focused commentary and reporting of significant events, also to more aptly meet the needs of our non-academic membership.

But, to return to my trip down Memory Lane. In the process of thinking about underutilized segments of the profession, it occurred to me that there were at least two other resources that fit this bill—our retired members and our disabled members. I felt I knew a good deal about the latter and I will discuss their situation later in the column. I asked the AAG office for information about our retired members and, on quickly receiving relevant data, began that unexpected trip.

Apparently about 45–50 AAG members transfer to the Retired Members category each year—at least recent evidence suggests this. There is a Retired Geographers Affinity Group with about 125 members. Robert Harper is the current chair and is only too willing to communicate with those who would like to affiliate. In glancing at recent retirees I was jolted time after time as I read the names of recent retirees who had played such an important part in my life as a geographer: Yi-Fu Tuan, Duane Marble, George Demko, Robert Harper, and Les King among many others in 1999; Jack Ives, Gordon Clark, David Hooson, and Alice Rechlin Perkins among others in 1998; Saul Cohen, Norman Thrower, Neil Salisbury, et al. in 1997; Emilio Casetti, John Kolars, Harold Rose, and Peirce Lewis et al. in

1996! Half of our National Academy members—Chauncy Harris, Gilbert White, John Borchert, and Waldo Tobler—are among the "active" retirees. They and many others (like Robert Kates) will never truly "retire" and will stalk the halls of academia as long as they are able. As I read the list of retirees, clear memories of the books and papers, the stories, and what may become geographic legends flooded my mind. Their individual and several impacts on my own professional development flashed before me.

So, how do we use this multitude of talent and expertise? Sad to say, we *under*utilize it. Of course we read and revere their past academic contributions—innovative contributions which are still being produced to this day—and we use their videotaped bio-sketches to give insights into their lives and thoughts. But what role do they play in Association affairs? In a professional association with only 7,000 members, how can we afford not to keep our retired members active in Association and disciplinary affairs?

All of us age. At a certain time in our lives, usually dictated by university or state policy, we retire. Now, what does "retirement" mean? For most academic retirees, it means cessation of undergraduate teaching, limited involvement with graduate students, and either a more or less complete break with our former academic pursuits so as to do things or go places that we previously had not done or visited. That's more or less a conventional view. But many of our retirees maintain an active interest in research, teaching, advising, departmental affairs, and Association affairs. This occurs despite the fact that little concerted action has been made by the profession to encourage and facilitate their continued activity. As one who sees retirement on a rapidly approaching horizon, I am becoming acutely aware of the lack of options open to a retired geographer who may not wish to fit the conventional mold.

My department's five-year plan makes it clear that I am at the top of the list of potential retirees. Consequently I have more than a passing interest in the relationships between the AAG and its retired members. At this stage the Affinity Group of Retired Members operates within the Association and organizes "informal" sessions at the AAG annual meetings. This year the AAG has received a request from its retired members Affinity Group to help them maintain an active interest and participation in professional geographic affairs. Considering that this group contains a valuable store of information and experiences regarding their systematic, regional and technical specialties that is somewhat unique in the discipline, and considering that the group includes half of our National Academy members, then it seems that there are resources here that the profession should continue to develop, access, and use. I draw the attention of all members who are planning to organize sessions at future annual and regional meetings of the Association, along with Program Committees for those meetings, to the possibility of using our retired members as a resource in many different scenarios. Are you looking for an expert for a panel discussion? Do you want an experienced and knowledgeable chair for an organized session? Do you need discussants whose knowledge of what has gone on in the discipline for a good part of this century is unequalled among most of our present membership? If so, contact Robert Harper, and he can send you a membership list together with an indication of the special interests of each member.

One of the goals of the AAG Director Ron Abler has been to develop a list of aware professionals whose names could be supplied to national, state, and local governments and other institutions and who could be called on as potential consultants, expert witnesses, or provide other specialist services on demand. Our retired members could form the backbone of such a group. And at certain times, such as the upcoming 100th anniversary of our discipline, it would make sense to draw on the memories and experiences of this group as part of our attempt to evaluate past achievements and practices and to help provide suggestions for defining areas of potential change.

Beside our retired constituency, there are other groups that are somewhat marginalized within the profession. The recent formation of a Disability Specialty Group draws our attention to the fact that little provision has been made over the years to ensure that our disabled members can participate fully in Association affairs.

Like most other sectors of the economy and society, both in the United States and more generally worldwide, perhaps the group that has been most marginalized and has attracted the least general support are disabled persons. That this has happened is not due to any deliberate conspiracy to subjugate or oppress this group by governments, businesses, educational institutions, or families, as is sometimes argued by some geographers who have recently directed their attention to this group. Rather, there has been worldwide recognition that some people have physical or mental functional limitations that mandate a different type of interaction with other humans and with environmental systems in which they have to exist. Humankind has progressed economically, socially, and politically without undue efforts to compensate for these differences. There is a tendency in some of today's literature to argue that disability is a social experience rather than a physical or functional one. But, as a young disabled female geographer from England has said, it makes no difference as to the social system in which you find yourself, a blind person still can't see, and a mobility aided person in a wheelchair can't easily negotiate a plowed field. In other words, regarding disability as being societally caused in the same way that ethnic, cultural, and religious groups or females have been disadvantaged is at best a very incomplete and not very productive way of looking at this problem.

So what has the AAG done to help integrate its disabled members into mainstream activities? This year at the Annual Meetings in Honolulu, some new practices were initiated. A Disabled Geographers Specialty Group was formed with aims of both encouraging research into the geographic consequences of disability and with intentions to act as a lobbying group for encouraging the AAG to institute more projects that assist its disabled members to participate more fully in Association activities. The AAG's reaction has been swift and sympathetic. In Honolulu, for those who are vision-impaired or print handicapped, tactual maps of the convention area were prepared and made freely available at the registration desk. A voluminous large-print version of the program also was made available, as was special equipment to enhance print so that programs, notices, and other printed communications could be accessed. Volunteer sighted guides were made available to take vision impaired or blind members to different paper sessions and to social events. For those relying on wheelchairs or other assistive devices for mobility, volunteers to push wheelchairs and help members to find important facilities ranging from meeting rooms to water fountains or bathrooms were available. In future years the AAG hopes to have floor plans of the convention area marked with the location of ramps and elevators so that this group will more effectively be able to plan attendance at different sessions and other organized events.

For those with hearing problems, sign language interpreters were made available—although in Honolulu, no one requested this service. In future years, starting with Pittsburgh, Program Committees and Local Arrangement Committees will be asked to take the needs of disabled persons into consideration to ensure the entire program is accessible. Situations where meeting rooms can be accessed only via stairs or obscurely located service elevators, will, we hope, not be part of the working profile of convention facilities. But to do this effectively, the AAG needs to have information on the number and types of disabilities present among our members. To this effect the next membership call will have a section which allows disabled persons to voluntarily indicate if they have a disability. A follow-up survey will try to determine the variety of disabilities. While not expecting a 100% indication of the presence and nature of disabilities among AAG members (which is of course very personal and private information that needs to be protected at all times), it would give the organization an idea of the extent to which it should plan for ways to mitigate obstacles and barriers faced by those with different forms of disability. One area, for example, that requires attention is that of incorporating disabled persons into field trips. Actions might be as simple as requesting that vehicles to be used on fieldtrips have mechanisms for loading and securing wheelchairs. In the more distant future we would also hope that vehicles will be equipped with infrared auditory transmitters that identify vehicle type, destination, boarding areas, and entrances and exits for vision impaired, blind, or print handicapped members.

These are but some of the many things that a professional organization can do to help its disabled members increase their level of participation. But there are also limits on what can be done. It seems that a considerable amount of experimentation will be needed at the next few national meetings of the AAG to explore the extent to which efforts can be undertaken to meet various equal opportunity goals. I suggest that organizers of the different regional meetings also give some thought to how they can facilitate the integration of disabled persons into their meetings.

By focusing on these groups, I do not mean to ignore the importance of other marginalized groups within our professional organization. In fact, I would hope that members of groups not mentioned here whose interests are not being met by AAG practices, will write to me and make suggestions as to how the AAG could alter its current practices to improve the use of all our human resources.

JANICE MONK: *Who's Teaching Whom? And How?*
April 2002

Two STORIES in the 2 March 2002 Arizona Daily Star are headlined, "Speaking Spanish, Texas Candidates Debate the Issues" and "It's Official:

Law Makes English Iowa's Official Language." The Texas debate between Democratic gubernatorial candidates is thought to be the first of its kind. It signals that politicians know the importance of changing how they try to reach the community. The Iowa bill is accompanied by one funding the teaching of English to immigrants—including those the state hopes to attract to fill employment vacancies and boost its dwindling population.

How do we, as teachers of geography, relate to these contrasting positions? If we are to become a profession that is more representative of the population and that reaches all learners, what strategies should we adopt? Should we change how we do our work or try to change others to be more like us? Or some of both?

Over the last three decades the AAG's membership has diversified, though more in its gender than in ethnic composition. Women constituted 7.4 percent of the college and university faculty membership category in 1973, 12.9 percent in 1985, and 24.1 percent in 2000. Data on degrees conferred in geography[1] show growth in women's representation (22 percent of BA/ BS and 7.6 percent of PhD degrees in 1973-74, 32.3 percent of BA/ BS and 24.1 percent PhDs in 1984-85, and 33.9 percent BA/ BS and 33.1 percent PhDs in 1996-97). Women made up 42.3% of the AAG's student membership in 2000. If most student members are in graduate programs, it seems that gains in women's participation have been greater at advanced levels than at the undergraduate level, and that growth at that level may no longer be occurring. Certainly, geography is not attracting women undergraduates in proportion to their representation among undergraduates as a whole, where they now make up more than half the student body nationally.

I hesitate to claim cause and effect, but it may not be coincidence that gains in the representation of women on geography faculties have also been greater both absolutely and relatively in the doctoral granting departments than in those awarding the bachelors as their highest degree. Julie Winkler documented this using 1997-98 data[2] and it was reiterated in the results of a survey of departments conducted by AAG intern Clionadh Raleigh, in 2001. Winkler showed that 35 BA/ BS geography departments had no woman faculty member, 33 had one only, and 14 had either two or three.

Gains in minority representation have been negligible. In 1973, the AAG estimated that minorities constituted 1.9 percent of the membership (128 Blacks and 30 Spanish Americans), by 1985, representation had risen to 6.4 percent, largely reflecting the inclusion of Asians in the minority category. Other minorities made up 2.2 percent of the membership. By 2000, minority membership was 8.3 percent, with Asians accounting for 5.1 percent of the total. Since a substantial number of the 325 Asian members in 2000 are likely to be foreign-born (and not all are working in the US), it's clear that US geography continues to fail in attracting other than the white population to any significant extent. Raleigh's survey data show that just under 10 percent of 1101 full-time faculty were minorities—77 men and 32 women. Minority students accounted for about 10 percent of students. It doesn't look as if we will see much increase in proportions of minority faculty in the near future unless concerted and systematic efforts are made.

So, what about teaching and its relationship to this picture of representation? Major introductory texts in human geography are finally making substantial efforts to integrate the material about women and gender that has been produced by three decades of research. We have also begun to see writing about the implications of different teaching styles in relation to multiculturalism and to sexism, homophobia, racism, ethnocentrism, and nationalism. I have summarized work on these themes in the *Journal of Geography in Higher Education*[3]; a recent issue of *International Research in Geographical and Environmental Education* takes up opportunities and hazards.[4] Materials prepared for Finding A Way, the K-12 curriculum and teacher development project on under-represented groups, directed by Rickie Sanders for the National Council for Geographic Education, contain much of relevance for college level teachers.

All evidence suggests that this is challenging work—whether the teacher is a caucasian or minority faculty member. As Audrey Kobayashi has remarked about teaching a course on race and racism, "I sometimes feel as if I am carrying a bomb into class."[5]

Another arena for action is the special programs designed to attract, support, and mentor minority students. According to Raleigh's report, before the AAG's COMGA project (1965-1976) one African American had a PhD in geography and three African American men taught in geography departments. When COMGA ceased, 42 PhDs had been, or were soon to be, awarded to African Americans. I recently asked on the AAG's listserve for geography department chairs whether departments were engaged in minority recruitment programs. From some 30 responses, I learned that most have no systematic departmental efforts, though some connect with institutional programs that, in a few cases, have provided meaningful support to minority graduate students in geography.

I would like to publicize some of the initiatives department chairs reported. Women students at several institutions have organized Supporting Women in Geography (SWIG) groups[6] to promote mentoring and professional development. Partly like the Iowa language legislation, they assist women to cope with the existing system, though by its existence, SWIG also challenges departmental cultures. Several programs support minorities in various ways. The University of Louisville has a scholarship for an African American student majoring in geography and a high school program that introduces African American (and other) students to GIS and information about careers, including local careers. West Chester University is reaching out to a largely minority magnet school in Philadelphia with the Advanced Placement geography course. Chicago State University, which is situated in a community that is 90 percent African American, has extensive commitments to service learning and internship programs in neighboring communities, has developed a strong student organization, and has also participated in a MacArthur Foundation-supported honors program. Mark Bouman comments that these students are motivated to "do something" for city youth. Helping them to do that with geography is a critical step. Darrell McDonald at Stephen F. Austin State University is working to engage American Indian professionals in east Texas in his programs with a view to developing a recruiting relationship. Another promising arena that I see is greater collaboration with community colleges that frequently enroll higher proportions of minority students than four-year colleges and universities. These approaches have in common building relationships as a route to enhancing minority participation.

To change geography's record with women and minorities we need to bring in critical perspectives (including analysis of whiteness), content that reflects the diversity of our communities, and diverse pedagogical styles, including those that connect with communities. We need to support and mentor students, to present geography as a field with career prospects, and as one that can contribute to enhancing the quality of life for all.

Notes

[1]Published in the Guide to Graduate Programs in Geography in the United States and Canada, 2000-2001.

[2]Julie A. Winkler, "Faculty Reappointment, Tenure, and Promotion: Barriers for Women," *The Professional Geographer*, 52(4), 2000, 737-50.

[3]Janice Monk, "Looking Out, Looking In: the 'Other' in the Journal of Geography in Higher Education," *Journal of Geography in Higher Education*, 24 (2), 2000, 163-177.

[4]*International Research in Geographical and Environmental Education*, 10(2), 2001, 168-201.

[5]Audrey Kobayashi, "'Race' and racism in the classroom: some thoughts on unexpected moments," *Journal of Geography* 98(4), 1999, 179-182.

[6]Arkansas; British Columbia; California-Los Angeles; Colorado-Boulder; North Carolina-Chapel Hill; Southwest Texas; Texas-Austin.

❋ ❋ ❋

Chapter 7

THE PROFESSIONAL SOCIETY

***

Without exception, every AAG President has used at least one column to examine the roles and functions of the professional organization. More often than not, these columns focused on two primary themes—the value of regional meetings (which many of the large PhD departments do not regularly attend), and the format of the annual meeting (more posters, plenary sessions, different formats). A sample of these columns follows.

***

I.M. MOODREE (GEORGE DEMKO)
April 1987

THE REGULAR COLUMN by the AAG President is being replaced this year by the letters from a friend of G. J. Demko. These letters provide insights about the profession from the perspective of a non-geographer. Letter number eight follows:

> *Dear George,*
> *Greetings from lofty Karakorum. This is indeed an exotic place and an interesting time—politically and climatologically. As Mr. Eliot so aptly put it:*
>
> > *April is the cruelest month, breeding*
> > *Lilacs out of the dead land, mixing*

Memory and desire, stirring
Dull roots with spring rain.

I note that this is also the month for your big annual bash. If I remember correctly the venue is Portland this year—a beautiful place indeed. The subject of your meetings, however, brings up the topic of what a professional society does for its members (or should do!)

You have educated me this year in matters of Association activity. Perhaps I should pass your ideas on to the geo-doers.

Clearly the annual meeting is the highlight of the year from both a social and intellectual standpoint. The program is a veritable bazaar of topics, ideas, etc. You will, however, remember my admonition about focusing the research interests a bit (I know your plenary sessions are aimed at that problem but I'm not convinced!).

I do think the Association publications are the most significant services rendered, which means that you should work hardest at maintaining and improving their quality and support level. You have a way to go on that count, but you seem to be making progress.

I fear most of your members perceive little other activity beyond these two functions. I hope your joint efforts with sister societies at geographic education are as vigorous and supportive as you state. Certainly the time is right (and will be short) to capitalize on America's concern with its geographic illiteracy.

I was surprised and interested to learn of other Association efforts such as working with the Consortium of Social Science Associations (COSSA) which lobbies for research funding. Impressive too are the representational activities for the profession at such organizations as the National Academy of Sciences, The American Association for the Advancement of Science, and other groups. A strong argument can be made for the fact that geography would not even exist in many institutions or in the minds of many leading intellectuals and policy makers if the Association were not publishing, sending reps and delegates, lobbying and making its presence felt.

Let me just remind you and your leaders of two things: 1) your geo-persons pay a goodly amount of dues and registration fees for these services and so expect them to be done well and their buck used effectively, and 2) your organization can always do better (and should continually try).

TERRY G. JORDAN: *Priorities*
August 1987

FOLLOWING ON THE HEELS of my two predecessors in the AAG presidency, both of whom were resident in Washington and heavily engaged in the real-world life of the Capitol, I must seem even more completely the cloistered scholar. I do not even know any acronyms. Carnival now yields to Lent, marketplace to ivory tower.

I am an academician. Perhaps my view is distorted, but I believe priorities in the AAG are dangerously confused. The membership of the AAG is, and always has been, composed preponderantly of professional academicians and students in higher education. The central purpose of such an association should be to promote and disseminate scholarly research in the discipline. All other activities in the profession—college teaching, K–12 instruction, and the various applied fields in geography—rest ultimately upon our ability to produce a substantial body of scholarship in the form of publications. Without this ever-expanding base of knowledge, we have nothing to teach, nothing to apply, nothing to justify our continued existence.

Budgets reflect priorities. I do not detect an adequate concern for research and publication in the AAG budget, either now or in the recent past. Consider the condition of our major scholarly journal, the *Annals*. In 1975, expenditures on the *Annals* amounted to $71,300, while income derived from the journal totaled $48,581, a net expenditure of $22,719. A decade later, in 1985, we passed a simply incredible milestone. In that year, the *Annals* yielded more income, $79,995, than we saw fit to spend on it, $79,894. The 1988 budget proposes to allocate only $74,500 (only $3,200 more that in 1975!) to the *Annals*, while anticipating $87,250 in income from it. Astoundingly, the editor was criticized for spending too much at the very time the *Annals* became a money maker. We have got a real jewel of a profit generator here, folks. Maybe we ought to go public with it on the commodities exchange.

I submit to you that something is dreadfully wrong with the prevailing budgetary priorities in the AAG. Any discipline that neglects to emphasize and fund its academic core is asking for trouble, and apparently getting it. We had better reorder our priorities. Then, maybe, we could revive the recently discontinued resource papers, pay for the ailing or aborted "spotlight" publications (which we now propose to fund out of annual meeting registration fees!), maybe publish another item or two in our moribund monograph series. Why, we might even quit bashing editors and let them improve the *Annals*. I asked Susan Hanson how she would have spent an extra $50,000, had it been available for the *Annals*, and she mentioned such items as better quality paper, adequate pay for the editorial assistant, map supplements, in-house cartographic work, and flexibility of length and content. Instead, we are starving our major journal and all other AAG publications. I hasten to add that the fault certainly does not lie with the Publications Committee, which, under Duke Winters' able leadership, has literally gotten blood out of a turnip.

We compare very unfavorably with other academic associations. The Society of American Archaeology, with almost exactly as many members as the AAG, for the fiscal year 1985-86 spent $95,222 on its major journal, *American Antiquity*, while deriving only $77,070 in income from it. On the annual membership renewal form, the Society guarantees that at least 50% of the $50 annual regular dues will go to fund publications. I have not heard of any doctoral-granting archaeology programs being abolished. Maybe we should be truthful and declare below the *Annals* masthead that "we don't spend a cent of your annual dues on this, your major scholarly publication, but instead steal from it like bandits to fund god-knows-what."

Perhaps another contrast is in order. The Geological Society of America, with just over 16,000 members, in 1985 spent $200,000 more on periodicals than it realized in income from them, and in addition devoted a net of $150,000 to other publications. In effect, $21.50 of each member's dues went directly to pay the net cost of scholarly publications. I have not heard of any doctoral-granting geology programs being abolished. We, by comparison, propose in our 1988 budget to spend $27,000 less on our scholarly publications than we derive in income from them.

Maybe mine is a minority view. It usually is. Please write to me, and I promise to present your collective viewpoint at the October council meeting, when the 1988 budget will be submitted for approval.

David Ward
August 1988

Over the past few years this column has addressed a variety of issues but the direction and thrust of the Association has naturally been its predominant focus. This focus has confronted the degree of emphasis the Association should place on its various functions. At times the merits of some of these functions have been judged to be undesirable extensions of a more narrowly conceived mission of the Association. Some time ago the Association undertook some new initiatives that were intended to consolidate and improve existing commitments to the world of geography beyond academia. The Association was expected to join with other organizations in providing stronger advocacy of the place of geography within educational enterprise as a whole and to reach out to geographic practitioners in both the public and private sectors. More recently concerns have been expressed about the degree to which this broad agenda is consistent with that critical role of the Association as a patron of research scholarship and especially its publication program. Over the course of the past two years, the publications committee has been responsive to these concerns, and the conduct of our annual meetings might benefit from a similar kind of creative attention.

I am convinced that however necessary it may be to confront these matters when we allocate our sparse resources, we should also emphasize that some key activities of the Association ought not to be viewed as mutually exclusive alternatives. The Association has a critical role in facilitating communication and mutual understanding of the various kinds of geographic education and of the diverse expressions of geographic interests and practices. We should however be cautious about expanding our efforts into activities for which we are unprepared and unqualified. Under these circumstances we ought to encourage the efforts of those fraternal organizations that devote their resources more directly to a larger and more diverse constituency. At the very least we need to continue to develop an effective and honest liaison with other geographical agencies and associations that may propose somewhat different views of geography than that professed by many members of the Association.

No doubt the debate about the direction and effectiveness of the Association will continue. Perhaps some common ground will be found as we begin to consider the prospect of a major participation in the meetings of the International Geographical Union to be held in Washington, DC in 1992. Already it is clear that an alliance of the broadest possible constituency will be necessary to ensure the kind of meeting that will project complex and imaginative views of geography. This meeting is both an opportunity to be the patron of an international research dialogue and a challenge for us to be sensitive to the connections between academic geography and its diverse and often latent projections in the community at large.

DAVID WARD
October 1988

OVER THE PAST DECADE specialty groups have become integral parts of the organization of the Association. The groups were initially devoted to the conventional range of systematic, regional, and methodological subdisciplines of geography but they now include a broader range of more precisely defined arenas of interest. Some groups support major journals, many distribute newsletters and almost all organize sessions at our national meetings. As part of the participatory fabric of the Association, they have served to enhance the networks of common interests and to improve the thematic structure of the annual meetings. Many have taken important initiatives connecting research in geography with that in related specialties of other disciplines.

At Phoenix many sessions were devoted to discussing the research traditions and future agendas of the Specialty groups. These deliberations will be published in a volume titled *Geography in America* edited by Cort Wilmott and Gary Gaile. This benchmark provides us with an opportunity to discuss ways in which Specialty groups might continue to be vigorous segments of the Association but also to enhance their connections with each other. At present there are 40 Specialty groups and it is no longer clear that they are of the same genre. They vary in size, purpose, and vitality. They began as a means to create discreet areas of discourse within a broadly defined discipline, but they may now project an image of an infinitely and peculiarly divided Association.

At Phoenix, different Specialty groups were clearly patrons of sessions devoted to similar if not identical issues and, unavoidably and inadvertently, some of these complementary sessions were scheduled at identical times. In the interest of developing connections within geography, it would seem appropriate that each Specialty Group sponsor joint or common sessions as well as individual sessions. Specialty groups, like curricula, record the incremental effects of changing methods and shifting interests. Periodically it is necessary to consider recombination and reformulation rather than supplementation and multiplication. Future program committees should perhaps identify broad research issues that might attract the collective attention of several Specialty groups. I sense a growing commitment to a more integrated vision of the discipline which at the very least should collapse some subdisciplinary walls. I hope that in formulating their programs for the Baltimore meetings, our Specialty groups will respond to this challenge.

DAVID WARD
December 1988

THIS ISSUE OF THE *NEWSLETTER* contains the program of our Annual Meeting to be held in Baltimore in March and it prompts me and I hope you to think about the content and the format of our proceedings. The Long Range Planning Committee concluded that "the open meeting with heavy reliance on specialty group-organized special sessions should continue, but that the meeting should be lengthened to lessen scheduling pressure and that an alternative means of program preparation should be developed to enable the highest quality, pre-screened papers to be featured in semi-plenary sessions." I am skeptical of the desirability or the need to lengthen the meetings, but I do believe that future programs would benefit from "semi-plenary sessions" which provide for longer presentations and co-ordinated commentaries. There were times during our recent meetings when the extremely low attendance at several sessions prompted the comment "everybody speaks but nobody listens."

We certainly need to devote a higher proportion of our proceedings to broad interpretative and controversial issues in order to balance a program necessarily dominated by the presentations of individual researchers. These general sessions with a broad sponsorship might also reach out to leading scholars in related disciplines. In this connection the Council has agreed to provide future program committees with a modest fund to which a group of at least three specialty groups might apply collectively to support a lecture by a distinguished scholar identified with a related discipline. These proposals would be part of an integrated program of several specialty groups which would be enhanced by fertile links with research developments in related disciplines.

Efforts to create a general program by means of "semi-plenary sessions" will certainly force the program committee to decline some submissions or to direct a larger proportion to poster sessions. The combination of open meetings and specialty group sponsorship has certainly invigorated the annual meetings but we do need to assess the impact of the proliferation of specialty groups on the agenda and program of our annual meeting. The identification of issues which connect the discipline needs the active patronage of the program committee. At the very least we should explore the possibility of an alternative format for at least some parts of a three-day conference before we accept the proposal to extend the length of our meetings.

JUDY OLSON: *Regional Division Meetings*
January 1996

I HAVE HAD THE OPPORTUNITY to attend five of our nine regional division meetings this year, and a good experience it has been!

I have seen the hills of South Dakota, tromped in the mud of a dolostone quarry in Minnesota, visited the unusual city of Oak Ridge, Tennessee, gazed at Lake Champlain, and heard interesting papers on such topics as the consequences of the (unfulfilled) prediction of reoccurrence of the New Madrid earthquake. I met numerous new colleagues in both the student and faculty categories and renewed various old acquaintances.

Our regional division meetings seem to have sometimes had second-class status over the years, at least in areas where PhD departments do not participate. In addition, the failure of some meetings to draw the desired crowds led to talk of merger of at least two divisions and to thoughts of rearrangement of boundaries. It is interesting to see what a few good ideas and some targeted work on the part of organizers can do, however. After rumors that the East Lakes Division was on the verge of extinction, the meeting this year had over 70 papers and drew upwards of 145 people to its business meeting. The work that Holly Myers-Jones put into the program, the supportive attitude of Division Chair Henry Moon, and one great idea from Rob Kent (hold the business meeting over lunch and include the price in registration) did wonders.

I think many people (not all, I am sure) would agree that regional division meetings are probably more important than ever at this point in history. The AAG annual meeting has become an extremely large event and shows no signs of scaling back—not that we want it to. Regional meetings, on the other hand, are much smaller and provide an opportunity to interact with a variety of colleagues, not just those in the sessions for our own areas of specialization.

Graduate students, especially doctoral candidates, are under increasing pressure to have professional presentations on their vitae before being considered for permanent jobs. AAG annual meetings are excellent opportunities, but so are the regional ones for all sorts of reasons. Ironically, the regional meetings are probably less intimidating, yet I have heard more questions and challenges from the audience (a less anonymous one) at regional meetings than at national ones, and questions come from a wider variety of listeners. Both the lower intimidation and the variety of questions are good experiences for graduate students.

Informality runs higher at regional meetings. Anyone with an idea for a good restaurant or watering hole is likely to be group leader for the event. And comfortable clothes are likely to be suitable for most organized activities.

Regional meetings tend to be inexpensive. Registration fees are modest; the division can choose among many venues because of the limited size of the conference; and with sites generally within driving distance, a university, business, or personal car or van can transport several attendees at shared costs.

The regional meetings are a place to innovate, to gain leadership experience and recognition, and to generate discussion back home. Anyone with a good idea for a session or field trip will probably get a welcome reception by the program committee. Likewise, for someone launching a new research project (including theses, dissertations, and funded research proposals), a clear presentation at the regional meeting can bring feedback while there is still time to make good use of it. Someone volunteering to chair a session will make the program committee happy. Division officers become good candidates for national committees and Council, and the Regional Division's Council representative is a good conduit for names to be carried to the national-level Committee on Committees, Nominating Committee, and other groups looking for people who have known they are interested and can get things done.

Field trips are also an advantage at regional meetings. Even if graduate students plan to take jobs far from their homes or alma maters, they will probably be expected to have some expertise on their region of origin. What better opportunity than a regional meeting to participate in field trips and learn more about our own backyards.

When several people have attended a conference, the experience makes a good departure point for faculty/student discussion back home as well. Other local geographers in attendance at the regional meeting might be invited to join the discussion. Participants can do a post critique of presentations, information gained, and anything else going on at the meeting. It is not only good fodder for interactions but a venue for instilling in graduate students (and ourselves) attitudes toward learning from others, communicating clearly with colleagues in our professional presentations, and offering service to ensure good meetings.

We faculty (and prospective faculty) can also be astute observers of the graduate students in our region. We often do not have time to hear students at national meetings, but the regional ones allow us to keep an eye on their interests and progress and either be ready to recruit them or pass on names to others in our network when they are ready for employment.

Several regional divisions lately have successfully combined meetings with State Alliances, and the proximity has been useful in spreading information about such important developments as Geography for Life. Sometimes these combined meetings tend to be two separate ones with the same venue instead of integrated ones, but even then it increases the chances of rubbing shoulders with interesting colleagues that we have little opportunity to meet otherwise.

Regional divisions have attracted a number of community college faculty and, in some cases, geographers who work for state and local governments and in the private sector. We could probably improve our connections considerably in those areas. Jobs for graduates, internship connections, and temporary employment for advanced graduate students are potential rewards for the four-year and grad departments, not to mention the intellectual stimulation provided by colleagues working in different environments than our own. These geographers, in turn, benefit by the same phenomena: new hires, interns, temporary help when needed, and the intellectual stimulation of colleagues from a different environment.

Perhaps the dual trends of globalization and localization portend good things for regional meetings. Globalization suggests we extend ourselves further spatially; localization suggests we pay close attention to what is going on at local scales. Regional meetings allow us to do that—we extend ourselves outside our own places of employment while becoming more intimately acquainted with our own regions.

What can we do to make better use of regional meetings in the coming years? First, consider participating in next year's meeting and put it on the calendar now. Give a paper, do a poster, suggest an innovative workshop, organize a session on a topic of interest to the region, recruit someone to organize a panel session on internships or on teaching prospective teachers or on jobs outside academia. Second, encourage others to take part in the meeting by notifying grad students and colleagues, encourage them to submit a paper, or pass on information to geographers in the area who might not be members. Some seemingly modest actions can make a big difference for regional meetings. Rob Kent probably spent only minutes passing on his good idea about combining the ELDAAG business meeting with a luncheon; what an effect it had!

PS: Being fond of giving quizzes, I offer this one on Regional Divisions: 1) Who is the current Chair of your home Division? 2) Describe the area included in your home Division or outline the boundaries on a map.

JUDY OLSON: *Annual Meeting Sessions*
April 1996

OUR NORMAL FORMAT at AAG Annual Meetings is the session with three to five presenters equally dividing the time. Papers are usually read and are often accompanied with slides or overheads or, in recent years, with images and animations produced with presentation software. Poster sessions have also been an important component of our meetings, with a core of visual images and an opportunity for one-on-one and small group interaction between observers and presenter.

A concept sweeping higher education is "active learning," and it is bound, sooner or later, to change the ways of annual meetings. Active learning, as the term suggests, has those on the receiving end actively doing something. It is difficult to attend events with an active learning format and not be convinced of the worthiness of the concept. Active learning is not altogether new, of course. In teaching we have long included lab exercises, "quizzes" not intended for evaluation, rhetorical questions, and Socratic method, all intended to stimulate active participation by learners. Such techniques are expanded and elevated in the current active learning movement, and they lead to more interesting and effective learning than straightforward, often dry, lectures.

How do we shift to a more active learning mode at annual meetings? Workshops in interactive format can be very effective but are not a likely substitute for papers. Poster sessions have the advantage that they come a long way toward active learning. The opportunities for interaction and the constant choices of posters increase our participatory level as an audience. We would do well to include more poster sessions, and we need to teach our students how to do poster presentations—how to handle them effectively as well as prepare the visual materials well. A good poster presentation, with worthy content and effective communication, is a satisfying intellectual course for the observer, and a good poster session a veritable feast.

There are other active learning models as well, six of which are described in the excerpt from *ASA's Footnotes* reprinted on page 12. Read them carefully and let me know what you think. Better yet, consider a poster format or one of these models next time you submit a paper or organize an AAG session.

Giving a paper seems to have become the reason for attending AAG meetings. Preparation of a paper unquestionably leads to learning on the part of the presenter. An active learning approach to presentation recognizes that the audience is, at that stage, in the position of learner. Success of presentation depends on how effectively the audience learns. Perhaps the balance of attention at presentation time will shift from presenter to audience as more active learning strategies are pursued.

LAWRENCE A. BROWN: *Continuity and Change: Geographical Societies: The mid - to late-1800s*
January 1997

SCIENCE BECAME A BROAD-BASED ETHOS in North America and Europe; confidence and excitement reigned; elements of transition from an agrarian to industrial society. The renaissance scientist was still the norm; specialization, and its narrowing scope, was decades in the future. Opportunities for geographical discovery abounded—the arctic poles, Africa, Asia, parts of Latin America—also the ocean, the lithosphere, the moon! And, while these endeavors gripped the fascination of many strata of society, there also emerged people who could afford to participate, directly or vicariously—wealthy philanthropists and a burgeoning middle class.

Geographical societies flourished, popularizing our science—Paris, 1821; Berlin, 1828; London, 1830; Mexico City, 1833; Rio de Janeiro, 1838; and in 1851, the American Geographical Society of New York. *Geographers in the Making: The American Geographical Society, 1851-1951* (J. Kirtland Wright, American Geographical Society, 1952) states:

> "The blossoming of geographical societies was part of a change in the whole aspect of Western civilization. After the Napoleonic Wars, the progress of scientific discovery, of technical invention, and of education caused a rapid increase in the quantity and variety of occupations and interests of all kinds . . . also of unions, societies,

and associations representing them . . . the railway, . . . steamship, . . . improvement of postal services and publication [facilities] . . . men were free to roam the world in mind and body. Energies . . . were now released for widespread commercial and colonial ventures . . . the settlement of remote frontiers . . . geographical discovery and exploration were their accompaniment" (pp. 7-8).

On a more personal note, *Geography in the Making* tells us that:

"Membership in the Society enables [one] to meet other gentlemen with kindred tastes, to listen to geographical lectures and take part in lively discussions . . . [George Knickerbocker notes] so far as one can tell from this map, the best route for the railroad would run from . . . the distances are so tremendous . . . he wonders whether the whole proposition is not an idle dream. But it doesn't seem so long ago that maps of his region were mostly blank . . . When he was a boy, the United States stopped at the Mississippi. Then came the Louisiana and Lewis and Clarke's wonderful journey . . . which step-by-step have filled in the blank spaces . . . he swells with pride . . . that the stars and stripes now wave over California" (pp. 8-9).

Also,

"On or near Washington Square stood other institutions of learning—the Union Theological Seminary, the General Theological Seminary, the new Free Academy (later . . . called the College of the City of New York), the New York Society Library" (p. 6).

This was a time of cultural-institution building—museums, libraries, colleges, theaters, parks, and learned societies. But it also was a time when organizations such as nineteenth-century geographical societies were motivated to urge

"the execution of geographical projects conceived for the benefit of the United States, from the building of the railways and the cutting of the Panama Canal, to the provision of adequate charts of aerial navigation." (M. Aurousseau, review of Geography in the Making, Geographical Review, nd).

Indeed, geographical societies and their principals were active players in, and often shaped, current events of the day; Isaiah Bowman is only one example. The brew-pots of geographical societies and US history intertwined.

The National Geographic Society emerged a quarter-century later, in 1888 (*The National Geographic Society: 100 Years of Adventure and Discovery*. C.D.B. Bryan, Harry N. Abrams, 1987). It had an egalitarian ideal,

"to increase and diffuse . . . geographical knowledge . . . [to] any interested citizen". (p.27) While early National Geographics were more scientific, detached, and with a tinge of judgment and polemic, it almost immediately moved toward the more popular format we know today (pp. 28-29)—tapping the market created by "emergence of a vast, educated, ambitious middle class (generated by the increasing sophisticated, expanding public-school systems and easier access to colleges and universities) . . . "(p.83).

Other nineteenth-century societies still existing include the Geographical Society of Philadelphia (1891, now "Greater Philadelphia") and Geographic Society of Chicago (1898). Though more modest than the AGS and NGS, these also supported exploration, sponsored travel excursions, held public lectures (e.g., "Adventures in the Arctic" by Shackleton, "Our Airplane Dash for the North Pole" by Amundsen), created library collections, and awarded Medals—to explorers such as Amundsen, Byrd, Cousteau, Glenn, Hillary, Peary; writers such as Michener; academic and institutional geographers such as Bowman, Colby, Goode, Grosvenor, Huntington, Lattimore, Salisbury.

New geographical societies appeared after World War II; among those on which I have information is the California Geographic Society. Founded in 1946, it is strongly oriented toward geographic education in K–12 and community college settings; these professionals dominate its membership, and the organization serves a large number of venues and people throughout the state. The Florida Society of Geographers was founded in 1964. Its concern with field trips and research on Florida has been broadened by close linkage with the Florida Geographic Alliance. Judging by the one meeting I serendipitously attended, this is a group of academic and applied geographers (and educators) from throughout the state, who value an opportunity to associate with one another. The New

Mexico Geographical Society, of more recent origin, largely involves a lecture series. Elliot McIntire writes:

> *"There is also the Los Angeles Geographical Society, founded ca. 1950, which is more like the 19th-century Geographical Societies. Most members are not professional geographers (although the officers are) but community members with in interest in travel, [field trips], and seeing slide shows of exotic places."*

Another organization is a class by itself; The Society of Woman Geographers, founded in 1925 by "four women . . . in New York—all recognized explorers" (Society Brochure). This national organization has seven local chapters in New York and Washington (the original two), Chicago, Los Angeles, Miami, San Francisco, and Seattle. It carries out an active fellowship program for "young women studying for advanced degrees" (>100); has awarded Gold Medals to Amelia Earhart, Jane Goodall, Mary Leakey, Margaret Mead, and Kathryn Sullivan; and "the Society's flag has been [carried by] active members [to] . . . new geographical horizons, including the depths of the ocean, the South Pole, and outer space".

Geographical societies are an interesting and vital part of our heritage. They emanated from a new age of science, exploration, and democratization of knowledge. These elements remain, but over time, travel, and education have moved to center stage. Especially important is the tremendous role of these societies in K–12 geography, including particularly the Geography Alliance and Geography for Life movements. The academy has been a constant, albeit changing, ingredient. Pre-World War II societies have been critiqued for presenting an ethnocentric view of the world, but they've been a vibrant and indispensable medium for popularizing geography. Geographical societies provide testimony that geography has been, and remains, exciting—for ourselves as professionals, but also for the public at large. Continuity and Change.

JANICE MONK: *Family Reunions and International Relations*
August 2001

SEVERAL TIMES RECENTLY I've heard people refer to the annual meeting of the Association as a "family reunion." Spending a few days with old friends, going to departmental parties, socializing at the opening reception, in the exhibit areas, hallways, and bars all add pleasure and meaning to our professional event. Annual meetings foster a sense of belonging and of continuity in our lives—like family reunions. But the family is not the same as it used to be. We now host substantial numbers of international "relations."

The June *Newsletter* reported that 25 years ago, the New York meeting had 1,706 paid participants of whom 129 hailed from outside the United States; of the 129, 114 were from Canada and five from the United Kingdom. Contrast that with this year's meeting in New York—4,764 participants, of whom 866 came from outside the United States. The percentage of international delegates has more than doubled in 25 years. The pattern was comparable in Pittsburgh, and my hunch is that the Boston, Fort Worth, and San Francisco meetings were similar. Canadian geographers no longer make up the largest contingent of international participants. An astounding number of British geographers—452—attended in 2001, over 100 of whom were students. When one considers that the annual meeting of the Royal Geographical Society with the Institute of British Geographers attracts an average of 900–1,000 registrants, the magnitude of British participation in our meetings is remarkable.

But the most notable change may well be in the attraction of the AAG for geographers from the rest of western Europe, Australia, and New Zealand. Among the Europeans, most evident in New York were German, Swedish, Finnish, Danish, Norwegian, Swiss, Italian, Dutch, and Spanish geographers, including students. Some of them regularly attend AAG meetings. Perhaps this change in participation is not surprising. The dominance of English as the language of scholarly communication, rapid growth in use of the Internet, and changes in air travel encourage "globalization" in our discipline. But our meetings are not really global. African, Latin American, Middle Eastern (other than Israeli), and Eastern European geographers were barely represented in New York. Only five of our Mexican neighbors were present. Among Asian geographers, most were from Japan or Hong Kong.

Do these numbers matter? Should we be doing anything about our "international relations?" How should the AAG and its members relate to the International Geographical Union (IGU), a long-established organization which holds regular congresses and conferences and has numerous thematic commissions? I've participated in the IGU since the 1970s; my impression is that United States and British geographers don't participate relative to other parts of the world and in proportion to their numbers.

And now western Europeans are increasingly turning to the AAG. Are we developing a discipline in which geographers from the wealthy nations write about globalization but mostly talk to each other?

And do we really talk to each other? Some European geographers have told me they feel isolated at the AAG meetings unless they have prior connections from graduate school or sabbatical experience. So we speak and they listen? The importance of "positionality" in the creation of knowledge has received considerable attention in recent theoretical writing in human geography. How might our work be enriched by dialogue with our international colleagues? German geographers, for example, conduct a considerable amount of research on the geography of the United States; their questions reveal some preoccupations not widely shared here. At the IGU Congress in Seoul in 2000 I was impressed that Indian geographers often concluded their papers by offering policy recommendations. We would more likely refer to the theoretical implications of our projects.

Getting better acquainted with those who come to AAG meetings would enhance our capacities for international collaborative research and education. Surely research on pressing environmental, political, economic, and social issues of our times would benefit from collaboration and dialogue with scholars who are based in other contexts and whose world views differ from ours. Educational activities, whether class projects or study-abroad ventures, could be enriched by contact with students as well as faculty outside the US.

The AAG has a Committee on International Research and Scholarly Exchange. I have asked its newly appointed chair, Gary Gaile, to develop and implement an agenda. The AAG office is considering steps to make our meetings more hospitable to international visitors and ways to give them more visibility. I will ask specialty groups to consider initiatives such as designating hosts at meetings, allocating resources to help widen representation, or sponsoring international collaborative projects. I encourage geographers who teach to look at the resources of the International Network for Teaching and Learning in Geography (*http://www2.glos.ac.uk/gdn/inlt/index.htm*).[1] For information about the IGU, its forthcoming congresses and symposia and its commissions and study groups check its website (*http://www.igu-net.org/*). I welcome your suggestions for action. Let's continue to enjoy our family reunions, but take in our international relations too. *Note:* [1] Ken Foote is the US coordinator for INLT.

ALEXANDER B. MURPHY: *What If There Were No AAG?*
April 2004

AS THE ASSOCIATION OF AMERICAN GEOGRAPHERS enters its second century, it may seem almost perverse to ask where we would be without the AAG. Raising the question may nonetheless help us gain some perspective on our organization. This is of particular importance at a time when the meaning of membership in any scholarly/professional society such as ours is undergoing some sea changes.

The AAG was founded in response to a perceived need for a society of geographical experts. In its early years, the society (with the help of the American Geographical Society) sponsored meetings and sought to nurture geographical scholarship. Within less than a decade, the *Annals of the Association of American Geographers* was launched and the size of the annual meeting expanded considerably. The AAG was thus set on a track that placed publications and the annual meeting at the core of the organization's activities.

The impact of these early developments was profound. To this day, the Association's journals and annual meetings rank at the top of the list of membership benefits in many geographers' minds. (In any given year, there is a strong correlation between the number of participants in the annual meeting and the overall number of AAG members.) The problem, of course, is that scholarly materials are increasingly available online at limited or no cost, and opportunities for participation in scholarly exchanges are proliferating as new interdisciplinary programs emerge and new forms of collaboration are fostered by the Internet. How does this leave the AAG?

"Vulnerable" might be the answer, if the AAG were only engaged in sending out journals and hosting meetings. Yet the AAG is much more than that. It is an organization that supports the production of geographical scholarship, not just its dissemination. It is an organization that gives geography a place at the table of disciplinary societies. It is an organization that provides links between universities and the public and private sectors. It is an organization that facilitates communication and interaction among geographers.

The AAG is not altogether alone in playing these roles, but I firmly believe that the discipline's position in American education and society would be immeasurably weakened without the AAG. With growing online access to published articles, the receipt of journals may seem less valuable than it once was. Yet without the AAG there would be no *Annals* or *Professional Geographer,* nor a *Newsletter* or Web site reporting on matters of discipline-wide concern and providing a forum for advertising position openings. The AAG annual meeting is certainly not the only place where one can practice and learn about geography, but without the AAG there would be no single moment when the entire breadth of our discipline is regularly brought together in one venue—with all that implies for intra-disciplinary dialogue and the cross-fertilization of ideas.

These things are just the tip of the iceberg, however. Consider a few hypothetical questions. Would geography's role in colleges and universities somehow be stronger if deans and provosts could not look to a scholarly society such as the AAG (or if they knew that such a scholarly society did not exist)? Would the profile of geography somehow be stronger in Washington, D.C., if there were no D.C.-based disciplinary organization working to promote the visibility of geography at the National Science Foundation, the National Research Council, and countless government agencies? Would the position of geography somehow be enhanced within the disciplinary matrix if geography did not have an organization with personnel who could meet and consult regularly with counterparts in other disciplinary organizations? Would geographers somehow have a wider range of opportunities if the discipline did not have a large scholarly/professional organization working to develop research partnerships or promote education and outreach initiatives?

Of course not—and those negative answers speak for themselves. When we join or renew our membership in the AAG, we are not simply paying to get a handful of publications or a discount rate at the annual meeting. We are making it possible for those publications and meetings to exist in the first place—and for an organization to exist that plays a crucial role in ensuring geography's continued vitality. I ask you to consider this the next time you are tempted not to renew your membership for a year because you might not be going to a meeting or because you don't feel the need to receive hard copies of the Association's publications. I ask you to bring it to the attention of colleagues the next time you hear them say they have no interest in the AAG because the meetings and journals devote insufficient attention to their particular interests. If you or your colleagues believe that it is important to enhance geography's profile in education and society, becoming an AAG member or renewing your membership should be a debatable proposition only if (barring economic hardship) you feel the hypothetical questions raised above can be answered in the affirmative.

None of this is meant to suggest that the AAG is doing everything right. We presumably all have ideas about what the AAG can and should do better—and the AAG should not obscure the larger picture. The AAG is vital to the health of the discipline, and every member who joins or renews makes it possible for the AAG to do more. Against this backdrop, our recent growth in membership is encouraging. If we build upon that trend, we will also be building a stronger discipline.

＊＊＊

Chapter 8

INTERNATIONAL RELATIONS AND ISSUES

❋ ❋ ❋

Regional geography is one of the long-standing traditions within the discipline and the decline in regional perspectives and international understanding is a focal point of many presidential columns. In this chapter, presidents speak to the importance of the discipline in strengthening international studies, geography's pivotal role in broadening society's view of the world, and relations between US geographers and those from abroad.

❋ ❋ ❋

RON ABLER: *Is Geography Losing Its Birthright?*
January 1986

THE 30 NOVEMBER *WASHINGTON POST* carried an announcement that White House science advisor George A. Keyworth was resigning to form a new company that will "show clients how to find and analyze political, economic, cultural and other information for use in foreign markets." Sounds like geography to me.

The 4 December issue of the Chronicle of Higher Education described efforts by individual states such as Florida, California, Oregon, and Texas to promote international education. A position paper written by Governors Bob Graham of Florida and Martha Layne Collins of Kentucky notes that citizens:

> *"should be able to speak with understanding the language of our clients, our allies, and our trading partners, as well as the lan-*

guages of our economic and political opponents. We must under-
stand the political and cultural systems that define their motivations
and interests as well as we do our own."

Sounds like geography to me.

International economic competition stimulated Keyworth's new venture and the decisions by states to promote the study of overseas areas. Keyworth is excited by the possibilities foreign competition opens up for teaching US firms about overseas markets. Individual states are concerned that their residents be sensitive to the needs of foreign investors and overseas trade partners.

All this sounds a lot like geography to me. Yet so far as I know, George Keyworth has no experience in the subject, his specialty prior to his White House appointment having been nuclear physics. The *Chronicle* article fails to mention geography as a discipline or even as a component of the programs being establishing in various states.

Geography arises, more than from any other source, from curiosity about places, about the linkages among places, and about the ways places affect each other. If geographers are not available to satisfy that growing curiosity or are not interested in doing so, others clearly will. At the same time that interest in overseas areas among the general public is on the rise, enrollments in regional geography courses continue to fall, and the discipline's corps of experts on overseas areas continues to shrink as foreign area specialists who retire are replaced by faculty with other specialties.

Teaching people about places is geography's birthright. Like any birthright, ours can be forfeited by failing to exercise it. If we do not move quickly to reassert our birthright, we may find it impossible to reclaim it from language departments, international studies centers, area studies programs, and private entrepreneurs.

The AAG Council has established an International Action Committee that is charged with taking immediate steps to augment the participation of geographers in international work of all kinds. The group will also formulate long-term plans for vigorous AAG promotion of international teaching and research. The International Action Committee is chaired by Dr. George Hoffman, acting secretary of the Woodrow Wilson Center's East European Program. It has eight additional members organized into subcommittees on international education, international scholarly exchanges, and collaborative cross-national research.

If you have specific suggestions regarding ways that geographers can insure that we retain our intellectual birthright, please share them with George Hoffman in care of the AAG Central Office.

SAUL B. COHEN: *Think Globally, Act Globally*
December 1989

GEOGRAPHY STANDS SECOND to none amongst the scholarly disciplines in its traditional concerns for global issues and studies of global systems. There would, therefore, seem too little need to use this column to remind AAG members of the importance of our commitment to maintaining and strengthening international links. However, in recent years there has been a widening gap between tradition and practice, and this discussion may be timely.

At the college curricular level, the emphasis on regional geography has progressively weakened at precisely the time of increasing attention to global studies programs in the secondary schools. At the graduate studies level, we have been party to more than a decade and a half of national disinterest in foreign area and language studies. If, as a result of the globalization of the American economy and the international political system, American universities reawaken to the need to reinvigorate foreign area studies, it is doubtful whether geographers would be able to play a vanguard role. We have paid far too little attention to training regional specialists in recent years.

I am not using this column as an exercise in hand-wringing, however. Some of our colleagues are deeply involved in interdisciplinary research and policy studies of the global environment—in the biosphere and the ecosphere in climate change, in global pollutants, in response to natural hazards. International political and international trade systems are receiving increasing, if still, limited attention. International development, too, is an important, if still too narrowly based geographical pursuit.

That there is still a broad base of interest to be tapped is demonstrated by an unprecedented response to the call for paper sessions at Toronto that will dovetail with the two plenaries. "Sharing the Global Village" and "Sharing the North American Continent" will each be followed by over forty sessions that specifically tie into these broad themes. This means that our membership has the capacity to respond to appropriate outlets that give attention to and voice internationally-oriented issues.

Another cause for optimism is in the opportunities to link up more effectively with the international community of geographers. Geographers in the Soviet Union and Eastern Europe have gained the freedom to interact with professional colleagues abroad, and both groups are seizing this opportunity. George Demko's work in linking Soviet and American geographers is a model for what vision and persistence can accomplish when political conditions permit. In responding to the heady challenge of how to develop meaningful ties with this part of the world, we should be mindful that Latin America and the developing world in Africa and Asia deserve greater attention.

For the near term, we face a particular challenge—China. You read in our last *Newsletter* of the Council's resolution condemning the repression of Tiananmen Square and its aftermath, but pledging ourselves to keeping open and enhancing contact with Chinese geographers. The time for action is now. The IGU Regional Meeting is scheduled for 13-20 August 1990. Planning at a time of funding uncertainty is not easy, but most of us are used to this. The US National Committee for the IGU is attempting to secure a sizable number of travel support grants. While current State Department policy does not encourage funding from federal sources, I believe that this attitude will be short-lived. The scholarly community as a whole, including the NAS/NRC, strongly endorses keeping communications open with China. For American geographers, the Beijing Meeting represents an opportunity to give tangible support to Chinese colleagues by their participation in the meetings, symposia and field trips. Those in the social sciences especially need this sort of encouragement.

International political uncertainties all too often sidetrack preparations and undermine important scholarly meetings. Let us keep Chinese politics in perspective by planning for a China of the summer of 1990, not 1989. This means taking the appropriate steps now to be part of a strong US geographical presence nine months hence. The benefits from becoming involved in International Geographical Union activities are hardly one-sided. We gain as individuals and as a geographical community when we extend our contacts and broaden research and teaching bases.

SAUL B. COHEN: *Seeking New Opportunities for Regional Geography*
March 1990

Disciplines tend to mirror the interests of their national societies in focusing upon particular kinds of problems and the methodologies most appropriate to analysis of those problems. Very often this happens as a response to cataclysmic events. After the Second World War, clinical psychology took on prominence because of the nation's concern with returning veterans. Later, as American society became consumer-oriented and concerned with self-gratification, clinical psychology not only came to dominate the general field of psychology, but shifted its base from academic and governmental institutions to private practice. Also, cultural anthropology shifted from its World War II-stimulated absorption with foreign cultures to including a major emphasis on indigenous rural and urban American groupings. This was in response to the rise of US domestic problems as reflected in the War on Poverty, the riots in urban ghettos, and the crisis of our cities.

The record for geography has followed a similar path—from the mobilization for intelligence and mapping issues during the Second World War, to the '50s focus on area studies in the context of America's global dominate position, to the urban/economic emphasis of the '60s and '70s, and the more recent turn to social and environmental issues. Today's focus in geography on data handling and data manipulation reflects the national emphasis on meeting the needs of the post-industrial, information age.

Now, cataclysmic events in Eastern Europe and the Soviet Union are gripping national attention. To be sure, globalization of the economy and the rise of Japan/Pacific Rim and the European Community have already made a strong impact on the American public. But the massive changes triggered by the collapse of Communism, attended by all-encompassing media coverage, have added a brand new dimension to our international consciousness. As the nation goes, so will the disciplines. Geography is not in the forefront of global studies in the schools, in foreign area analysis, or in the fields of economic trade and tourism, but we can and should do more. On the positive side, our leadership in areas of international development (the Bellagio Declaration on World Hunger, convened by Akin Mabogunje and Robert Kates, is an important and dramatic sample of what geographers can do) demonstrates geography's capacity to assume major responsibilities.

For far too many years geography has been bogged down in the debate over the "scientific" validity and utility of area studies. The argument that such studies are idiographic, not nomothetic (and therefore "unscientific") has become an excuse for intellectual inactivity. One only has to look to the recent volume edited by Gary Gaile and Cort Willmott (*Geog-*

raphy in America, Columbus: Merrill, 1989, 840 pp.) to see how limited is the research interest of American geographers in regional studies. (This volume is, I believe, a most useful contribution to the field and its editors and the authors are to be applauded.) The section that deals with international understanding points to the scant work in many African geography subfields, the decreasing numbers of specialists in geography of North and South America, to serious gaps in studies of Asia, and to almost no activity in Eastern Europe (the USSR being an exception). The volume even lacks a section on Western Europe. George Hoffman, our leading East European scholar, has been a voice in the wilderness with his concerns for attracting a new generation of geographers to his area, and Robert Harper has sought for a good number of years to stimulate the interest of geographers in developing regional studies for the school curriculum.

I am now pleased to inform you of a first step that the AAG is taking to put regional studies back on the agenda. We are proposing a joint US/Soviet Geography School project. The overall objective is to provide youngsters from each country with greater understanding of the geography of the other's country, through an issue and process oriented regional geography. The project will be carried out by a joint team of geographers and teachers, and be oriented to primary, intermediate, and high school levels.

A preliminary meeting that I convened in December with AAG, National Geographic Society (NGS), and National Science Foundation (NSF) representatives has yielded encouraging results. Since then, NGS has awarded us seed monies to support initial planning sessions. To launch the effort, George Demko, now director of Dartmouth's Nelson Rockefeller Institute of International Studies, and I will visit our Soviet colleagues to discuss next steps. We plan to organize a joint planning and working session in the US in the spring.

In implementing this project, we expect to involve our NCGE colleagues—classroom teaching strategies and teaching experiences must be an integral part of curriculum development.

Also under consideration are similar institutes for "paired" projects with Japan/the Pacific Rim and the European community. While we are presently targeting schools, I hope that we can ultimately extend the effort to the college level.

One final note—I have been encouraged by the responses that several of our members have made to some of the issues raised in recent columns. In a forthcoming column I'll share their ideas with you. Meanwhile, your reactions continue to be welcomed.

THOMAS J. WILBANKS
January 1993

CLEARLY, RIGHT NOW most of us are preoccupied with our own survival and growth as islands of geography within our individual institutions, but one of the things we are supposed to personify as a discipline is an awareness of the rest of the world. How aware, in fact, are we of the capabilities, activities, and working conditions of our fellow geographers in Africa, Asia, Eastern and Central Europe, Latin America and the Caribbean, and the Middle East?

In September, I was in New Delhi in connection with a new research project, and I visited geographers at Jawaharlal Nehru University (JNU) at the invitation of Dr. Saraswati Raju, who had attended the August International Geographical Congress in Washington, D.C. At JNU, geographers are grouped with economists and demographers in a Center for the Study of Regional Development, an arrangement that is even more unusual in India than in the United States; and they are bright, ambitious, well-trained, and hard-working. Let me share with you a few thoughts stimulated by that muggy afternoon, which was punctuated at the end by a monsoon downpour.

First of all, it seems to me that we geographers in the United States and Canada know too little about geography in the developing countries. Are there fellow geographers studying urban processes or environmental issues or geographic information systems in Santiago, Kuala Lumpur, Ibadan, Cairo, or Ljubljana? People with whom we share research interests and with whom we might share research data and results? People who hunger for interaction with the international scholarly community? I would be willing to bet that, whatever your own personal research specialty, geography departments in developing countries include several first-rate intellects whose specialties overlap yours and that interaction with them would benefit you intellectually as well as philanthropically. Too often, when it comes time for us to suggest possible developing country members of new working groups of the International Geographical

Union (IGU), for example, we come up blank, because the world is so incomplete.

Secondly, it is clear that in most cases our colleagues in developing countries are operating with entirely inadequate resources. Consider trying to do your jobs in the American Southeast in the summer without air conditioning, with painfully limited library resources (especially international journals), with few or no micro-computers, with annual salaries below those of your graduate assistants, with travel support entirely dependent on the beneficence of distant and often mysterious international sources. Under these kinds of conditions, many of our colleagues in other countries are performing heroically with too little recognition of their efforts. Are there ways we can help, even as our own resources are stretched ever thinner?

Why, indeed, should we give this particular issue a high priority when there are so many other demands on us? I would argue that there are many reasons, including moral ones. But maybe the most compelling for many of us is that, at the same time we are being asked to help our society understand the rest of the world better, we must recognize that most foreign area research by American geographers in the future will be through multi-national collaboration: individuals or teams from this country working with partners from other countries, both because our colleagues there can contribute understandings that we lack and because the days of academic colonialism are over. Several current projects with our Russian colleagues will be valuable models as we adjust to this new internalization of geographical research. In any event, without host-country partners, many of our proposals for foreign-area research support are going to be turned down, regardless of their intrinsic intellectual merit, and I have been supporting this kind of policy in the US federal agencies in which I work. We need to know geographers in other parts of the world, because without them we will know too little about those areas, because without them we may be unable to work there professionally ourselves, and because without them we are mining their reality for our own professional benefit without giving much in return.

Can we do anything to improve this situation, whatever our motivations? Not a lot in the near term, unfortunately; but we can start by getting better informed, and we can start to show leadership in sharing our fairly meager resources. For instance:

1. *The US National Committee for the IGU.* As a continuing member of this committee, I will be urging the IGU to commission an inventory of geography in developing countries and countries in transition, working through national commissions as appropriate, to tell us more about who is doing what, where, and under what conditions. Many of us will find colleagues with whom to exchange research information if someone will help us find them.

2. *AAG regional specialty groups.* Similarly, it seems to me that these specialty groups, who should be our windows to their respective parts of the world, should consider preparing summaries for the rest of us on geography in their regions: departments, faculty, topical interests, and particular potentials and needs. What does American geography need to know about its counterparts in that region, and how might we help?

3. *Resource sharing.* Finally, there are more tangible ways that we can start to make a difference, several of them already pioneered by geography in the US. For example:

    •   *Travel grants support international meetings.* Thanks to the remarkable fund-raising success of Tony de Souza and the solid stewardship of Don Vermeer, the 1992 International Geographical Congress set a new standard in supporting the participation by scholars from developing countries—younger professionals as well as entrenched senior figures—in such an international meeting. Tony and Don found that raising money for this purpose was relatively easy, if one really tried. We might try more often.

    •   *Sharing publications.* As a good example of what we can do, the AAG supports the AAAS Sub-Saharan Africa Journal Distribution Program, which provides subscriptions to nearly 200 journals in the sciences and humanities— including the *Annals of the AAG*— to about 200 university and research libraries in 38 countries, identified in collaboration with regional organizations. I hope geographers involved in other journal publication efforts will encourage more sharing of this sort, extended to other developing countries and regions besides Africa alone (Burma? Albania? Tadzhikistan? Cuba?).

- *Seek research collaboration.* As we learn more about colleagues in developing countries who share our research interests, we might be more active in proposing joint research activities and seeking funding for them. A promising international team effort focused on important issues is a good prospect for support, even these days, and we have far better access to sources of funds on our end than do our developing country colleagues. Remember, though, that collaboration means just that: joint development of ideas and a full partnership in carrying out the research, not exploitation for the sake of a foreign field research opportunity.

- *Adopt a department.* As we learn even more, and especially as we develop stronger personal ties to geography departments in developing countries, we might consider proposing sister-department relationships where their department's orientation is congenial and its location is related to our department's long-term regional priorities. Over a decade or two, staff exchanges might be arranged; students might move in both directions for training and field experience; information and materials might be shared; and each partner would be shaped by a sustained pattern of interaction in some depth.

In these kinds of ways and others, I think it is time for us to be more active in reaching out to the rest of the world at the same time that we work to consolidate our new status in North America. Not only is it the right thing to do; but it can add excitement and learning to our enterprise and, in the spirit of the 1992 IGC, help us to realize anew that "Geography Is Discovery" in terms both professional and personal.

JANICE MONK: *Enlarging World Views In and Through Geography*
January 2001

WRITING THIS NOVEMBER COLUMN amidst the sorrows and apprehensions of late September is a difficult task. We mourn the loss of so many lives in the US on 11 September—people of diverse ages, nationalities, and walks of life, some of them involved in our discipline. We are grateful for the condolences that came from around the world. My mail included messages from colleagues in Azerbaijan, Turkey, India, Spain, New Zealand, England, and Israel, among others. Countless entries have been posted on listservs by geographers seeking resources to help their students understand the context for the attack, alternative ways in which it might be understood, and to support responses that would promote justice, equity, and peace.[1]

Given the uncertainties, I've chosen to focus my remarks on aspects of our professional practices and possibilities that I believe would enhance our prospects for making a difference in world affairs by enlarging our own geographic understandings and that would improve our communication with others.

What is the current status of US scholarship and teaching on global processes in relation to specific places? Texts and courses on world geography have often divided the world into discrete regions, paying relatively little attention to the processes that link places and create and reflect inequalities within and among them. Globalization has recently become a favored theme across the social sciences and humanities, especially in the research being promoted by influential bodies such as the Social Science Research Council and the Rockefeller Foundation. But too often, work focuses on macro-scale processes at the expense of in-depth understanding of the specifics. Conversely, area studies have been criticized for immersion in the specifics at the expense of the wider picture. As geographers, we need to speak up for studies of the global processes that are grounded in empirical studies of the histories, cultures, economics, and politics of specific places.

But how well are we doing in sustaining the necessary regional and area specialties, in fostering field studies, language competence, and communication and collaboration with geographers from other regions as complements to our other rapidly developing technical skills? In the present context, it's noteworthy that only in the last year has a specialty group in Middle Eastern geography been established within the Association, and it is tiny. Its creation represents a positive step, as does the initiation in 1998 of *The Arab World Geographer,* which includes authors from the US and abroad. Nevertheless, in only a few academic centers in the US are geographers being prepared to develop expertise on this critical region and on its external relationships.

Two recent articles draw attention to the dominant use of the English language and of authors and editorial board members from English-speaking countries in geographic journals that purport to be international.[2] A study measuring international co-authorship over the decade 1985–1995 found little such work in geography journals.[3] These findings prompt questions about limits on the circulation of knowledge, on who works with whom, and especially on whose perspectives are widely shared. The studies also remind us that commonly used databases are not globally inclusive. On a related issue, geographers in several countries have told me of the pressures on them to publish in the US or UK in order to be valued at home. The challenges they face are not only linguistic. Intellectual homogenization is fostered by expectations of reviewers from the dominant countries that specific authors be cited and theoretical positions invoked. We need to be more open to alternative perspectives and ways of thinking, especially in the face of theories that argue for the significance of plural voices and of "difference."

Deepening understanding of other places and perspectives can be attained in multiple ways. Personally, I've found that participation in the International Geographical Union, its congresses and commissions, has initiated friendships that resulted in exchanges of faculty and students, collaborative writing, and repeated or extended visits to colleagues' homes and departments. We may not all have extensive possibilities for travel, but electronic communication and the emergence of such groups as the International Network for Teaching and Learning in Geography (INLT) offer avenues for increasing communication. I am particularly interested in INLT's "Global Geographic Inquiry Challenge" which will link undergraduate geography majors and faculty across national borders in collective inquiries into "real world" issues. Its first module on migration will shortly be available.[4] We also need to explore approaches to teaching that engage students in examining the complex implications of values such as nationalism and patriotism, and of the stereotyping of those who are different from themselves within the US and beyond. Such teaching can engage students in reflection on their own values and behaviors in addition to extending their awareness of others' views.

Finally, I want to encourage greater engagement in international collaborative research. I am delighted to see that the AAG is co-sponsoring a conference this month at California State University, Northridge that brings together US and Chinese scholars to foster pan-Pacific cooperation.[5] Working together is not easy, and we need to pay more attention in our research and graduate education to the process involved in the creation of collaborative and equitable partnerships. Researching and teaching *with*, not only *about*, brings new dimensions into geographic understanding. Swiss colleagues have taken a lead in articulating principles for research partnerships that address such issues as building mutual trust, sharing responsibility, and creating transparency.[6]

In advocating that we strive to create and disseminate geographical work that expands world views by being more inclusive and collaborative, I don't expect that we will solve the global dilemmas and crises of our times. I do urge us, however, to reflect on how our practices might be strengthened in the search to bring our discipline to bear on attainment of the elusive goals of a more peaceful, just, and equitable world.

Notes
[1]See, for example, *http://www.peoplesgeography.org*; the Association of Pacific Coast Geographers has committed to including resources on its website *http://www.csus.edu/apcg*.

[2]Gutiérrez, J. and P. López-Nieva, "Are International Journals of Human Geography Really International?" Progress in Human Geography, 25(1), 2001, 53-69; Short. J.R. et al., "Cultural Globalization, Global English, and Geography Journals," The Professional Geographer, 53(1), 2001, 1-11.

[3]Abraham, I. and T. Harris. "Measuring International Collaboration," in International Scholarly Collaboration: Lessons from the Past. New York: Social Science Research Council, 2000.

[4]Contact *Michael.Solem@swt.edu*

[5]*http://www.csun.edu/~hfgeg003/*

[6]Swiss Commission for Research in Developing Countries, Guidelines for Research in Partnership with Developing Countries: 11 Principles. Bern: KPFE, 2000.*http://www.kfpe.ch/*

M. DUANE NELLIS: *One Year Later: Geography's Role in International Education*
September 2002

You should receive this column approximately one year after the tragic events of 11 September 2001. This single, terrible day crystallized the challenges of globalization and the importance of international research and education to our national security and for a more peaceful world.

Many have commented to me on how the United States has changed since this event, but a key question now is how the country is adapting to these changes.

This past November, then AAG President Jan Monk focused on "enlarging world views in and through geography." I would like to offer further comments on this general theme on the anniversary of 11 September 2001, especially as it relates to the importance of geography in international education.

The global transformations that have intensified in the past decade have created an unparalleled need in the United States for expanded international knowledge and skills. This need for enhanced international education has certainly been a theme in the geography community since the development of our discipline. That need gained considerable momentum during the late 1980s and 1990s through programs and efforts designed to highlight the importance of geography to the education system and to the general public, including such publications as the National Geography Standards document, *Geography for Life,* and the National Academy of Sciences report entitled *Rediscovering Geography.* Certainly it is clear to those of us in the discipline that geography is central to science and society in offering a distinctive and integrating set of perspectives through which to view the world around us.

From various indicators, including the disconnection between the United States as a superpower and the abysmal general knowledge of world geography, it is clear that our nation needs significant educational reform including adoption of geography as a core discipline. Understanding the global system through geography is prerequisite to interpreting everything from information affecting national security to acquiring the skills and understanding needed to foster improved relations with all regions of the world. Geography and foreign languages (among other subjects) are central to the ability to function effectively in other cultural environments and value systems, whether conducting business, implementing development projects, or carrying out diplomatic missions. Geographers must continue to position themselves at the core of critical global issues. The efforts coordinated by Doug Richardson, Thomas Wilbanks, and Susan Cutter on the *Geography of Terrorism,* for example, were timely and have brought geography national visibility by highlighting the key roles geographers can play in addressing such international issues.

Creating true international competence through geography will require substantial further education reform and financing, and partnerships among educational institutions and agencies and the private sector. Some progress has been made on this front, but I fear geography has lost some of the momentum attained in the mid-1990s. Geography is a relatively small discipline and developing productive partnerships will require renewed efforts on the part of many academic geographers as well as those in business and government.

The International Policy Paper *Beyond September 11: A Comprehensive National Policy on International Education,* published recently by the American Council on Education (2002), recommends national efforts that focus on the following outcomes relevant to the geography community. First, geographers need to significantly expand the international knowledge of faculty and graduate students in professional and technical specialties. We should be working with our colleagues in these fields to institute curricular reforms that enhance international education in subjects such as business. Second, geographers should make strenuous efforts to increase the diversity of United States students who major in international studies and enter international service or geography teaching. Third, geographers should provide government agencies, corporations, and the media with international geographic knowledge through geography departments and resource centers. Geographers should focus on placing major international events in their broader geographical, historical, and cultural contexts to enrich the primarily political perspectives in which they are currently portrayed.

Geographers also need to continue to prepare students to be more effective citizens in an increasingly diverse and multicultural society. Geographers should continue to strive to ensure that all students are exposed through the primary and secondary levels to the international and comparative content that is the core of the discipline. Geographers in university settings need to ensure that every postsecondary undergraduate experiences a curriculum that imparts cross-cultural understanding, an international perspective, and a broad acquaintance with major global issues and processes. Geographers should also encourage more students to study abroad. In many cases geography already plays a central role in such efforts. The challenge for us all is to continue to augment the contributions of geographers to international learning in the post–11 September world.

Certainly other disciplines have roles to play in global understanding and we need to partner with such disciplines to strengthen international education. We also need to promote and build on our partnerships with the K–12 community. The Geography Alliance network has made tremendous strides in the K–12 community but much remains to be done. As many of you are aware, geography is often buried within a general social science curriculum that sometimes leaves out critical elements related to global awareness.

Quality research on a broad range of international issues and effective methods for teaching geography are fundamental to raising global awareness. Government agencies and private foundations need to make a renewed effort to fund such research. Whether researching global change or the diverse roles of women in rural societies, geographers have much to offer relative to global understanding.

Geography should continue to be instrumental in facilitating the use of spatial technologies. Geographic information sciences and remote sensing offer students and non-geographers a sense of the importance of spatial sciences, and their use continues to expand dramatically in international business and scientific arenas as well as in military applications.

Geography is important to expanding worldviews and on the first anniversary of 11 September, we must reflect on our work as geographers and how we can increase our efforts to improve international education in the United States. The time for action is now.

ALEXANDER B. MURPHY: *Rethinking the Place of Regional Geography*
October 2003

EACH DAY BRING NEWS of momentous events unfolding in far-flung corners of the world: the emergence of North Korea as a nuclear power, crises of governance in West Africa, power struggles in Southwest Asia, ethnic conflict in Israel/Palestine. The causes and consequences of these events are varied, but we cannot even begin to understand them without the benefit of geographical analysis. Nonetheless, geographers are largely absent from public debates about their nature and implications.

Why is this the case? Partial answers can be found in matters well known to readers of the *AAG Newsletter*: the comparatively small size of the discipline of geography, the trivialization of geography in the public imagination, the absence of geography programs in some institutions of higher education with disproportionate influence in the public arena.

Yet geography cannot solely blame the outside world, for the discipline itself bears some responsibility for this state of affairs. A number of points might be made in this regard, but surely an important one concerns the marginal status of regional geography in the United States today.

To some, regional geography connotes an encyclopedic march through regional facts, and of course it was the rejection of that type of geography that led to a major shift away from regional approaches in the 1960s and 1970s. Unfortunately, our discipline moved away from regional geography without adequate consideration of either what might constitute good regional geography or what is lost when geography programs fail to produce strong regionalists. In practical terms, this has undermined the discipline's ability to contribute to discussions about developments in different parts of the world, has limited geography's involvement with communities of scholars and practitioners focused on regional issues, and has worked against the expansion of geography programs in colleges and universities.

Turning to the first of these practical matters, over the past two years Afghanistan and Iraq have loomed large on the international scene. Yet when a student comes to me and asks for a reading list of geographical works focused on these countries and their regional setting, I am hard pressed to come up with more than a small handful of (sometimes outdated) publications. As far as I know, the number of American geographers who have done any serious fieldwork in Afghanistan or Iraq can be counted on the fingers of one hand. Under the circumstances, it is no surprise that geography is rarely looked to as a source of information or insight.

More broadly, the complexities of the globe are so great that, for better or for worse, discussions among specialists are often organized along regional lines. Meetings of Africanists, Europeanists, and the like bring together individuals with a strong grounding in the languages, histories, and political economies of different parts of the world. "Geographies" should be part of this list, but are often missing—both because members of other disciplines rarely think in geographical terms and because few geographers define their areas of expertise in regional terms. Hence, geography and the perspectives of geographers are notably underrepresented in regionally oriented organizations and debates.

Finally, by turning away from regional offerings, geography programs have deprived themselves of a powerful opportunity for growth. My perspective on this matter may be biased by particular experiences at my home institution; but if the University of Oregon is at all representative, it is clear that careful nurturing of regional geography can play a role in a department's success. Our introductory regional courses are among the more popular offerings on campus, attracting strong students and leading many of them to consider geography as a major. Moreover, many of our regional courses have come to play a vital role in area and regional study programs at the university, winning friends for the department and making geography indispensable to a range of academic programs. Finally, our areas of regional emphasis have helped us attract strong graduate students who have developed a focused interest in a region of faculty expertise. I am convinced that a serious investment in regional geography could pay similar dividends in other places—and indeed in some cases it already has.

The discipline will not benefit from reengagement with regional geography if such an initiative is handled simply by organizing a perfunctory introductory regional survey course or by marching students through a set of regional facts devoid of concepts or meaning. Instead, our goal should be to produce students who have a sophisticated understanding of different parts of the world and their relationships to other parts of the world. This means focusing a critical geographic eye on regions: looking for explanations as well as descriptions, and situating regions in the context of developments at different scales and across time. It also means encouraging students to acquire the language skills, historical background, and field experience that is critical to in-depth regional understanding. The task is not a small one, but it is critical if geography is to play a more significant role in the national and international arenas.

* * *

Chapter 9

DISCIPLINARY DIRECTIONS

✳ ✳ ✳

Some of the most provocative and intellectually stimulating columns written over the past three decades centered on where geography as a discipline should be headed. Debates surrounding specialization versus synthesis, new directions, and future geographies in a changing world are explored.

✳ ✳ ✳

I.M. MOODREE (GEORGE DEMKO)
January 1987

THE REGULAR COLUMN by the AAG President is being replaced this year by letters from a friend of G. J. Demko. These letters provide insights about the profession from the perspective off a non-geographer. Letter #5 follows:

> *Dear George,*
> *Best wishes to your and your merry band of geogrumpers for the New Year! The new year here doesn't begin until 29 January and so I contend with the year of the Hare until then. Remembering your love of solitude and Eastern philosophies I was about to recommend you to the local monastery after your presidential term. I think, however, you have already done adequate penance for a lifetime of sins.*
> *You do realize that in thirteen more years we will cross over to the 21st century. Wouldn't it be worthwhile to look ahead and do*

*something interesting as a profession? Why not propose that your as-
sociation develop a volume of papers which will predict spatial/geo-
graphic patterns and trends into the new century? Instead of your
dull inventories and prospects and hoary guidelines, why not try to
speculate? What will the world's urban system evolve into by the year
2000? What political alignments or realignments can be anticipated
over this period? Will the Sun and Frost Belt shifts over space continue in the US, and what spatial patterns will result? Any number
of fascinating explorations in the geography of the year 2000 can be
made. And, why be concerned about absolute accuracy? Economists
are always way off the mark with their projections and they keep getting asked for more! After all, if geography is a science and an art, you
should be able to predict and describe with skill.*

*The Great Middle Kingdom is a most fascinating and remark-
able part of the world. It must certainly be a geographer's delight with
its array of languages, ethnic groups, physical landscapes and more.
Certainly this country has generated more innovations and absorbed
more than one can imagine. Has anyone written a comprehensive
historical geography of this country?*

*Keep my mail coming. Some of your readers certainly do have
thin skins or else some of my barbs are hitting close to the target.
The fact that the mail has been running about 10 to 1 in favor of my
wisdom greatly heartens me. I am sending answers to the most cantankerous, but can't respond to all.*

*I must continue my search for black jade and trace the loca-
tions of Van Gulik's Judge Dee stories. I recommend them to your
members (by the way, each one starts with a map!). Thus, I wend
my way east toward the drylands. I leave you with some old Chinese
wisdom—"all good ideas were once heresy."*

SUSAN HANSON: *On Geographers and the Environment*
May 1991

IT's NO SECRET that the environment is a hot topic. From TV spots
and grocery bags offering 25 handy tips on how YOU can help save the
earth (e.g., by using baking soda, vinegar, and boiling water instead of
Drano to clear clogged drains) to umbrellas sporting world maps and the
institutionalization of a fourth R (Reading, Riting, Rithmatic, and Recycling), awareness of and concern about an "environmental crisis" appears

to be at an all-time high. Across the country, public opinion polls show environmental issues at or near the top of people's concerns, courses with "environment" in the title are bursting at the seams, and plans are afoot to launch a National Institute for the Environment (NIE) to support environmental research and training in much the same way that NIMH supports research and training in mental health.

The hot topic of the 1990s, and perhaps just a passing public fancy, environment has always been at the heart of geography. Geography's long and rich tradition of focusing on people–environment relations puts geographers at the center of contemporary environmental debates and at the forefront of environmental research, teaching, and policy making in the coming decades. Geographers bring to the study of environmental issues a synthetic, integrative view and an ease with complexity that others often lack. Non-geographers increasingly recognize the special contributions that geographers offer: we are, for example, disproportionately represented on national and international committees on human dimensions of global environmental change, such as those established by the SSRC and the NAS/NRC.

The current blossoming of attention to the environment opens challenges and opportunities for geographers. We have the opportunity to conceptualize environmental geography more broadly as the study of people-environment relations, not simply as investigations informing environmental policy. Expanding the catchy "man–land" phrase to "nature–society" did more than admit women as active agents in shaping (solid) environments: the label change invited a rethinking of the focus and scope of environmental geography. As geographers, we are uniquely well equipped to explore the links between environment and economic restructuring, between environment and urbanization, and between environment and questions about the sustainability of life on this planet. We are uniquely prepared to see the connections between actions at one scale and impacts at another. The many subfields of environmental geography should draw together, not fragment, in this effort to understand the multifaceted people–environment puzzle. If we do so, we shall be able to play a leading role in the emerging environmental centers and institutes on our campuses. The educational challenge involves more than adding another R to the elementary school curriculum. We must ensure that our students learn about both physical systems and human systems, as well as interactions between the two. Too often we are content to educate students as "physical geographers" or "human geographers," but to immerse students in one side of the discipline to the total neglect of the other is to do them a disservice.

The limelight on the environment puts an essential part of our geographic heritage in the public eye, giving us another opportunity not only to clarify for others what geography is all about but also to contribute substantively to intellectual discourse and to creative social change.

Susan Hanson: *On Staying Connected*
June/July 1991

Carol Gilligan (*In a Different Voice*) would say it's because I'm a woman; others might say it's because I'm a mother, a feminist, or an academic, but I think it's because I'm a geographer that connectedness is important to me. Or maybe it's not because I'm a geographer, but why I pursued geography in the first place: because I delight in seeing connections where others don't.

The theme of the Miami meetings, "Local Lives and Global Systems," highlighted connections between and among different geographic scales: this theme reverberated throughout many papers and sessions. What connects Inuit lands in Northern Canada with my kids listening to the Beatles in Worcester? What links land degradation in Inner Mongolia with the cashmere sweater you buy from Land's End? What connects a wealthy white Westchester household with a black, female-headed household in the South Bronx? Geographers illuminated these and countless other connections in Miami.

Sameness and similarity are often the basis for connection (remember trying to define a uniform region?), and veterans of *Sesame Street* may find themselves humming "One of these things is not like the others" as they reflect upon the many and varied bases for finding similarity. At the same time, one of the key lessons of geography is that connection often rests on difference and, in fact, can create and sustain difference (remember complementarity in spatial interaction models?). Holding on to connections that bridge differences appears to be growing in importance for both intellectual and political reasons. Contrary to predictions that global economic and social integration would yield a homogenization of people and cultures (the ubiquitous McDonald's), integration now appears to be making difference more, not less important. In addition, geographers are learning that what might have appeared as homogeneous (e.g., women's employment patterns), actually masked a multiplicity of difference. We are, therefore, increasingly called upon to see, appreciate, and cultivate the connections among disparate places, processes, and peoples. We are increasingly asked to see the links between the self and the "other." The role of the "other" as different now draws a great deal of attention in debates about community building, about political alignments, about the academic canon.

Geographers' fascination with connections—our ability to see links that others cannot—makes us feel more acutely the inevitable tension between the desire or the need to specialize and the will to integrate, the dedication to not losing sight of the connections. At a meeting in Miami of the newly formed "Miami group" of young scholars (successors to the "Phoenix group" formed in 1988), one clear theme to emerge was that it was precisely geography's integrative dimension (integrating the physical and the human environments, integrating phenomena over different geographic scales) that attracted this generation of scholars to our field. Another theme, voiced again and again as a concern, was the intellectual tension between specialization and synthesis or integration. (Can one specialize in integration?) This group of recent PhDs in geography shared a reluctance to cut the web of connectedness that defines life on this planet, yet felt compelled to do some snipping in order to carve out "manageable" research foci.

I believe our unwillingness to chop reality into discrete but manageable bites is a source of our strength. The problems facing humanity now require illuminating the connections that link different places, different scales, and different peoples. I believe individual geographers can develop expertise in soils, fluvial processes, Bangladeshi agriculture, urban housing, or suburban retailing without losing sight of the connections that bind a particular problem area to others. Specialization and integration need not be mutually exclusive. Our ability to see connectedness where others may see only difference, and through that our ability to remain connected intellectually to our colleagues of many and varied interests, allows us to contribute to ongoing debates both within and outside of geography.

THOMAS J. WILBANKS
October 1992

FIVE HUNDRED YEARS AGO this month, Columbus landed in what we now call the Americas, and the mythology of this encounter is an indelible part of the education of most of us. For many, 1492 is the most memorable date between the time of Jesus and the year 1776. Many of us recall the names of the three Columbian ships more easily than the names of the first five or six US presidents. Our seat of government is called the District of Columbia, and other familiar imprints on our maps include Columbus, Ohio; British Columbia; and Colombia.

What we have found out, however, is that we were told only part of the story. Columbus was not the only explorer of the Atlantic, and his observations can only be considered "discoveries" from the painfully limited points of view of new travelers to lands that were already long-settled. Certainly, 1992 cannot be a celebration of the decimation of Native American populations from Alaska to Chile or of our blindness to the accomplishments of societies that were here long before the Europeans. Instead, it seems to me, the quincentenary should be considered at least partly a reminder of some challenges still before us.

One of the challenges, of course, has been to fill a number of glaring gaps in our scholarship about the pre- and post-Columbian experience; and in this regard our fellow geographers from Carl Sauer to Karl Butzer have been leaders in unraveling the truth. The special September 1992 issue of the *Annals of the AAG* is a significant addition to this record—a superb editorial initiative on the part of the AAG.

But another challenge is to learn from our experience as we look forward to another century (or two, or five) . . . Let me suggest several lessons that we might be able to draw from our experience with the Columbian encounter as we revisit it this year:

1. Humility about how much we know. How is it that we were taught so much that was wrong? Is it that scholars were caught up in "the persistence of error" (James, *Annals*, 1967)? Is it that the uncertainties at the frontiers of scholarship were poorly communicated to the frontiers of education in the schools? Even if "the art of teaching is careful lying" (*Teaching as a Subversive Activity*), because teachers so seldom have time to tell the whole truth, centuries of teachers got the story seriously wrong while scholars toiled away at their specialized pursuits. What stories are being told in our schools right now that are similarly wrong? Why?
2. Appreciation of different perspectives on the same reality. Recent research shows us how one-sided our views have been of the encounter

between European and Native American populations. What Europeans saw as wilderness, Native Americans saw as home. What Europeans saw as hostile, Native Americans saw as nurturing. What Europeans saw as a frontier of opportunity, Native Americans saw as a zone of encroachment. This, it seems to me, should suggest to us lessons about openness and tolerance to the perspectives of others, not just as a political necessity but as an intellectual opportunity. These lessons apply not only to our openness to alternative theoretical constructs but also to our interactions with non-scholars and non-geographers. Remember, for instance, the vastly under-appreciated insights of Gunnar Olsson about differences between what we think we are saying, as we work to get our messages right for ourselves, and what our readers may be understanding from the very same words (see *Birds in Egg*). We still have a lot to learn about the value of perspectives that lie outside our usual frames of reference—whether they are from other countries, other cultures, diverse users of our research, our students, or our fellow citizens.

3. Remembering the importance of field observations to discovery. Somehow, it seems to me, in this world of human-modified environments, mountains of quantitative data, and a shortage of funding for foreign-area research, we have lost touch with some of our traditions of direct observation, not only as a learning opportunity, but as a source of inspiration and joy. As the recent International Geographical Congress indicated, "geography is discovery" not only through analyses of library materials and data bases but also through encounters with the world through our personal experience. Karl Butzer suggests that these encounters often challenge our familiar ways of thinking. In some cases, in fact, formal training can even impose blinders on observation—an anti-intellectual outcome from an intellectual enterprise. The Columbian encounter reminds us both of the importance of field work in geographic learning and of the ways that field work, broadly defined, encourages an open mind about the complexities of the world around us.

Other lessons could be mentioned as well, such as the tendency for innovations to occur in waves rather than in isolation (Cabot and others in the same time period as Columbus) and the lasting impact of certain apparently random occurrences, such as the naming of America (in 1507 by a group of cartographers revising Ptolemy's world atlas in a work session at St. Dié des Vosges, France—which this month will host its third annual Festival International de Géographie).

These comments barely scratch the surface of what we might learn from our reconsideration of the Columbian encounter 500 years later, looking forward as well as backward. I am interested myself, for example, in Butzer's suggestion that the ecological impacts of the European arrival were a prelude to the global environmental changes introduced by the Industrial Revolution.

At any rate, let's try to take this special month as a time to rethink who we are and where we are going, avoiding both the traps of too much hubris ("Columbus discovered America") and too much humility (geographers only talking to small professional cohorts). The process of geographic discovery continues, affecting a world far beyond geography alone, and we are indeed fortunate to be a part of it.

LARRY BROWN: *The G in GIS—Getting It Right*
September 1996
GEOGRAPHY HAS REALIZED significant forward progress in recent years—in attention from other social and physical sciences, in the business community, in government, in education at all levels. AAG membership is at an all time high; several programs are expanding. It is, indeed, a very rosy picture. But . . .

The renaissance is attributable to many factors. The quality of our scholarship, teaching, and outreach is vital, of course. But we also must recognize the role of geography's academic traditions and strengths, which dovetail nicely with societal concerns about the environment, educational quality, globalization, and internationalization of the economy. Likewise crucial are organizational initiatives such as the Geography Alliances. GIS also belongs in this list, but I place it in a separate, arguably unique category. Why? Because it is a technological innovation that geography has played a major role in pioneering, developing, diffusing—at the forefront of the GIS revolution.

How has GIS impacted our profession? NSF's initiative towards a National Center for Geographic Information Analysis (NCGIA) is a prime exhibit. Judy Olson's Presidential Plenary Session at the 1996 AAG Meeting, Is GIS Killing Cartography? evokes a more ambiguous, but powerful portrayal of impact. GIS is the AAG's largest specialty group by a considerable margin; nearly every department of geography has GIS capability. In government and business GIS has become a central ingredient of environmental management, urban planning, facility location, marketing, transportation logistics, and the like. There are numerous GIS software and analysis companies, such as ESRI, and these mesh with new highly successful venues such as the Business Geographics Conference, *Business Geographics*, and *GIS World* magazines. With this backdrop, is it merely coincidental that a geographer is at the helm of the US Census 2000 effort? In our sister social and physical sciences, GIS is widely recognized and embraced, but more critically, this has opened the door to appreciating what geographers do and the significant contributions that derive from a spatial perspective.

In short, GIS's impact is nothing less than profound. But what if the acronym were SIS, for Spatial Information Systems? Almost certainly, the benefits accruing to us would be considerably less because, I believe, there is a strong serendipitous element at work, one related to association with the G in GIS. Not all gains fall under serendipity, of course, and SIS also would have serendipitous effects. But the G in GIS does make a significant difference, one from which geography has benefited immensely.

This raises a vital question. What should geography do, now and in the future, to get the "G" in "GIS" right so as to position ourselves for the next epoch—to leverage and build our present advantage such that GIS gains are not simply a wave that passes?

Though not a GIS expert, I've had to respond to the question in several settings. As a framework for doing so, geographers involved with GIS cite three types of endeavors—Routine-Descriptive GIS, using GIS software to make maps, diagrams, and the like; Analytical-System Design GIS, joining GIS with statistics, cartography, information retrieval, and similar tools to answer substantive questions of a scholarly or applied nature; and Technical-System Development GIS, advancing GIS software, analytic systems, etc.

Where along this continuum is the next frontier, waiting to be pushed out? My thoughts follow, not as dictum, but with the intention of opening discussion that is critical to our future.

Routine-Descriptive GIS capabilities, coupled with cartography, are important elements of our teaching and training mission. Due to ready accessibility and wide-spread usage, however, this domain is no longer geography's alone. As Joan Treadwell, Editor of *Business Geographics*, says (July/August 96, page 8) "Products are becoming more user-friendly and less costly at the same time. Access to the technology isn't reserved for specialists any longer." Routine-Descriptive GIS, then, is integral to our calling, will continue to hold a place in the curriculum, but represents an area that's been well explored.

Technical-System Development GIS has been the cutting edge area, and probably the focus of most GIS research in recent years. We share the domain with computer science, engineering, and other technical disciplines who carry out large parts of the endeavor and may do so even more in the future. Geographers need to maintain a role in developing spatial analytic GIS tools, a role that ultimately fuses with analytic cartography. This will require that some departments maintain the skills and critical mass to contribute. Technical-System Development GIS is an important aspect of our discipline; input of geographers is a significant component of the overall effort, it has been integral to our present success, and more will be done. Geography does not, however, enjoy a comparative advantage here, and a base for broadly building our future lies elsewhere.

Analytical-System Design GIS is, in my opinion, the locus of opportunity. This involves addressing important substantive problems and providing meaningful, insightful, trenchant answers that draw on the geographer's craft and reveal the power of our analytic perspective. The foundation, still in its infancy, is creating analytical protocols that link GIS with statistics, cartography, mathematics, modeling, information retrieval systems, et cetera—for ready use by geographers and others. Much work is needed here. Another aspect is carrying out substantive research, for which topical expertise is paramount, eclipsing GIS *per se*. This endeavor is even more in its infancy, with few examples of published work that advances substantive knowledge in key areas, using GIS. Moving ahead here will build on progress in creating analytical protocols and integrating them into the suite of readily available, widely used research methods. With our spatial perspective and kitbag of tools, long familiar to us but new to many others, geography has a comparative advantage in Analytical-System Design GIS. If done well, which must include a firm grounding in substantive issues, Analytical-System Design can be the next frontier from which we impact other sciences, business, government—and parlay geography's present position into enhanced strength.

A parallel can be drawn with geography's "quantitative revolution". In the 1960s and '70s departmental stature could be enhanced solely by technical expertise in statistical and mathematical analysis of geographic phenomena. Over time, these capabilities became more common, but also were augmented by packaged programs, highly accessible software suites, and routinized analytical protocols. Technical expertise became less integral; focus shifted more strongly to the research question and substantive or conceptual relevance. In the GIS revolution, the last phase of this transition is already under way; hence my emphasis on analytical protocols, their application, and the indispensability of a substantive frame of reference. There is a difference, however. Because quantitative methods are not inherently in the domain of geography, serendipitous gains are rare. GIS, on the other hand, has the G. Accordingly, as long as geography is at the epicenter of effective, path-breaking evolution we will gain stature, eminence, and visibility.

Will geography move to capitalize on our GIS gains? Or will we sit on the bubble while it's here, in complacent bliss? Tune in.

LARRY BROWN: *Continuity and Change*
October 1996

I RECENTLY WAS GIVEN "THE EDUCATION OF A GEOGRAPHER," Carl Sauer's address to the AAG as Honorary President in 1956. Reading this reminded me that contemporary issues often are re-inventions of, or a re-visitation to, the past. This is true of academic scholarly issues, also of administrative institutional ones. Progress is made, new frontiers are honed out, but in doing so we often forget our predecessors. It is a fine line, of course, between "old wine in new bottles" and "new wine in new bottles," a line often in the mind of the beholder. An important component of moving forward is revolution in the literal meaning of the word, denigrating what is left behind. Ultimately, however, there comes a collecting together of new and old, a period of consolidation and synthesis wherein we recognize our roots, embrace the continuity of scholarship, and of the academy. Human geography, I believe, is in the midst of this latter, synthesis phase of the dialect of science.

This moment in geography cannot be captured, or even sketched, in one President's Column. I do, however, provide two illustrations of the dialectic of science and the ongoing–present synthesis in human geography, drawing on Sauer's assertions.

One interesting aspect of Sauer's article is the tension between topical-systematic and regional geography. I was first introduced to the dichotomy as a Northwestern graduate student, where we were required to have both a regional and topical specialty. Over the years, emphasis on either has waxed and waned, depending on the time, department, journal, etc. Currently I am involved in a broad-scale effort concerning the nature of international activities within university settings. Once again, the dichotomy arises—area specialty versus systematic knowledge. The pendulum now seems to have shifted toward the latter. Symptomatic is an announcement for the International Dissertation Field Research Fellowships of The Social Science Research Council and American Council of Learned Societies. After many years of area-based programs, this now is concerned with

> *"scholarship that treats place and setting in relation to global and transnational phenomena as well as particular histories and cultures . . . [T]o learn more about [an] area means . . . how that area is situated in events and processes going on outside its culture or economy or ecology . . . [A]n integrated understanding of transnational and global phenomena . . . cannot be acquired without reference to specific places which give shape and substance to those larger processes."*

In one sense, this is *déjà vu* in that Sauer addresses a similar tension, with a similar conclusion. He opts for topical courses which "have the advantage that they are analytic, and their elements may be scrutinized at any scale of inspection and by more or less adequate techniques."

Concerning regional geography, he notes

> *"I do not accept the idea that anyone can do the geography of a region, or comparative geography, when he knows less about anything he assembles than others do . . . the ineptly named holistic doctrine leaves me unmoved; it has produced compilations where we have needed inquiries . . . if [we] stayed on the trail of themes rather than of regions our contributions to knowledge would be more numerous and of a higher order."*

But the regional geography of today is quite different from that of Sauer's era, and the SSRC-ACLS resolution reflects that transition. Steps along the way in geography include the excursion of quantitative geography into regional studies (ridiculed by many initially); locality studies; global change initiatives that focus on spatially-differentiated impacts, nationally–locally, of socio-economic restructuring; the concept of spatial contingent outcomes that emanate from broad, general forces; the new regional geography, etc. Hence, the *vu* of Sauer is not the *vu* of today; in between have been a series of synthesis, or dialectics, wherein there is movement beyond the past that incorporates the past.

A second observation is that even though Sauer speaks strongly of systematic geography, as noted above, he also opts for geography as art rather than science.

> "*A geographer is any competent amateur . . . of whatever is geographic . . . Really good regional geography is finely representational art, and creative art is not circumscribed by pattern or method . . . Is not . . . the harmonious landscape also something proper to think upon?*"

This indication of tension in Sauer's thinking between geography as art and geography as science, echoes in broader disciplinary concerns. In Sauer's case, and perhaps more generally, one element of these seemingly contrary convictions is an appreciation for field work, exploration, intellectual discovery, and original thinking or insight—which in its highest form does become an art. Hence,

> "*geography is first of all knowledge gained by observation . . . I like to think of any young field group as on a journey of discovery, not as a surveying party . . . One of the finest experiences . . . is to go where none . . . has been . . . to make some sense out of what has not been known.*"

But eschewing a scientific approach means Sauer,

> "*does not see our future in a retraction within limits that set us apart from other disciplines . . . We still stand uncommitted, though we are being advised that we too should have a properly defined methodology . . . we remain in a measure undelimited and . . . unreduced to a specific discipline. This is our nature . . . our present weakness and potential strength . . . What geography is is determined by what geographers have worked at, everywhere and at all times.*"

The echo here is of course with past and current debates concerning the role of scientific method and positivism, the perspective from which phenomena should be viewed, what is valid geographic endeavor (is this really geography?), how students should be trained, and the like. Hence, the simple dichotomy of art versus science has been elaborated many times through methodological-conceptual shifts such as quantification, spatial analysis, Marxism, social theory, post-modernism. So again, the *vu* of Sauer is not the *vu* of today; instead, there has been a series of syntheses, or dialectics—movement beyond the past that incorporates the past. And the dialectic of science goes on.

The pertinence of these two illustrations—systematic versus regional geography, geography as art or science—is enhanced by Sauer's stature as a major social science scholar from the early twentieth century onward, and his continuing influence on geography today. This stature also underscores a personal feeling. We too often eschew the past (or geography done in the way of the past) as part of justifying, or supporting, the present. Practically, this may be equivalent to "throwing out the baby with the bathwater." And personally, even in the dialectic's thesis-antithesis phase when denigration is most functional, there is no reason why compassion, understanding, appreciation, and respect for what has gone before should not be the order of the day.

In future columns and otherwise, I hope to touch more on the theme of continuity and change in geography.

PAT GOBER: *Compartmentalization of Geography*
August 1997

FRESH FROM MY ELECTION as Vice President of the AAG, I changed the way I approached the 1996 annual meetings in Charlotte. Instead of attending the full complement of Population Specialty Group sessions as I usually do, I decided to sample a broad array of fields. I went from Population to Industrial (now Economic), Urban, Political, and Geomorphology Specialty Group sessions. What I saw and heard were highly specialized groups of scholars honed in on very specific research questions, using their own vocabularies and methodologies. For the most part, I felt like I was on another planet. I realized, for the first time, that most of us do not really attend the same AAG meeting. We attend a meeting within a meeting focused on the activities of 50 or 100 other individuals whose research questions, methodologies, and vocabularies mirror our own. While theoretically there are opportunities for the cross-fertilization of ideas across specialties, I question whether many such exchanges actually take place.

The high level of specialization in our field challenges the very raison d'être of our disciplinary journals. Dating from my earliest years on the AAG Council, I have heard editors of the AAG *Annals* and *The Professional Geographer* lament the small number of submissions and the fierce competition from an ever-larger list of specialty journals. In a field the size of geography (the AAG has 7,300 members), the *Annals* receives between 100 and 150 manuscripts a year, in part, because most of us are more interested in reaching an audience of specialists than we are in communicating our ideas to nonspecialist geographers.

This type of thinking is also common in our geography departments. I heard recently that the large introductory physical geography course in one of our premier departments is now team-taught because several faculty members do not have the expertise to teach outside their very limited subspecialties. This is in a freshman level course! It is increasingly difficult to entice graduate student to take course work and seminars outside of their areas of specialization. As a result, seminars get smaller, the opportunity to hear fresh, outside perspectives declines, and the focus of graduate education gets narrower.

While I appreciate that geography is in no way unique in this regard, there are dangers, in my mind, associated with the growing compartmentalization of the discipline. Part of the success of and clamor for geography courses and content in recent years is the recognition of geography's integrative nature. Geography's strength in integrative knowledge and its marriage of natural and social science position it well to deal with increased national and global concerns about the interactions among human behavior, social institutions, and the natural environment. To realize this potential requires research and instruction that cuts across theoretical, methodological, and topical subareas. Future success demands that we be able to communicate with one another and with persons from related fields. A highly fragmented discipline and a philosophy of graduate training that emphasizes narrow specialties are not conductive to the process of intellectual synthesis.

Increased specialization, along with the increased diversity of our membership, also creates organizational challenges for the AAG. Occasions for the membership to convene as a community of scholars are in rapid decline. Each year the annual banquet draws fewer members, the business meeting is more a progress report to a small circle of diehard loyalists than it is a meeting of the minds on issues facing the discipline, and field trip attendance is down. Any sense of common values, the importance of long-standing traditions, and a shared vision of the future is in short supply.

There are a number of things we can do organizationally and individually to build a stronger sense of community without undermining the well-deserved success of our specialty groups or the logical specialization of our members. Starting in Boston, specialty groups will be asked to organize "state-of-the-art" lectures that will summarize their work for non-specialists. The AAG's upcoming Centennial celebration in 2004 is an opportunity to commemorate the discipline's past achievements, but more importantly, to chart its future course. Networking and leadership-building activities for new faculty members would also facilitate cross-specialty exchange and the sense of belonging to a larger community of geographers.

As individuals, we can design and offer courses that attract interest from a wide range of audiences, do joint teaching efforts, and insert graduate students into situations that broaden their backgrounds. We can also choose to send our best work to our disciplinary journals and balance the desire to attend specialty sessions with the need to stay abreast of broader issues in the field. For my part, I plan to stick with a more eclectic model of meeting attendance.

PAT GOBER: *Compartmentalization of Geography II*
January 1998

ONE OF THE REWARDING ASPECTS of writing the monthly president's column is the volume and the intensity of feedback that I have received. Responses to my August column about the compartmentalization of geography and the follow-up September column about the narrow nature of PhD education in geography were numerous, supportive, and often passionate. Your feedback has deepened my understanding of the problems that we face and has strengthened my resolve to address them. I would like to share with you in this month's column four themes that emerged in your communications to me.

First, many geographers, particularly senior-level individuals, bemoaned the lack of collegiality, community, and disciplinary focus at AAG meetings and in AAG publications. They fondly remember a more collegial and communicative profession in which members considered themselves geographers first and specialists second. Many people believe that the sense of geography as the glue that holds us together is on the wane. One person eloquently wrote that whenever we meet as geographers, the geography in us should be evident through our maps, our sense of space and place, and our feel for the connection among different scales. Unfortunately, this often is not the case.

Second, my characterization of the narrowness, inflexibility, and single-mindedness of current graduate education struck a chord among young geographers. Many graduate students are deeply frustrated at being forced into narrow specialties when what drew them to geography in the first place was its breadth and the opportunity for synthesis. A student in one of our premier PhD programs said:

> *"I was not prepared for the tunnel vision of the academic world. I would like to somehow combine my interests in research with applications to real world problems—an activity viewed with curiosity and suspicion within the geography academy."*

The geography academy should celebrate and cultivate, not discourage such interests. Several students told me that they are resigned to playing the academic game but resent being trained for highly-specialized research positions even when they aspire to teaching positions and careers in applied research.

A third theme involves the problem of communication. One byproduct of specialization and fragmentation is the development of what one person called "non-interchangeable dialects." A climatologist who regularly attends AAG sessions outside of his specialty has a problem "with presenters who show a complete lack of sensitivity to their audience." This comment echoes my own experience. I sense a perverse arrogance in making ideas so complicated and obtuse that they can be understood by only a few. If you lose all your listeners, you won't be accused of being a fool. Your listeners will be made to feel like they're the fools.

A recent article in *Science* speaks to similar communications problems in physics. Apparently, physicists have long fretted over their inability to communicate with the lay public. Now leaders in the field are worried that physicists can no longer understand each other. A group of reform-minded physicists and journal editors met recently to discuss the fog of poor writing and ways to clear it. Written guidelines will be presented to the publications board of the American Institute of Physics. These guidelines suggest that journal editors make clarity of presentation an explicit condition of acceptance of an article, that abstracts be made more generally intelligible and that the best-written articles receive special recognition by the journals. There is also talk of a new electronic publication, tentatively called *Highlights*, which would report on selected journal articles in a form comprehensible to physicists in any specialty.

A fourth theme running through my correspondence is the pervasiveness of fragmentation and specialization in science and society. A German colleague noted that "when you go to a session in industrial geography or in geomorphology or in quantitative methods (in Germany) you are in most cases no longer of the opinion that the people presenting papers are geographers."

Last month I attended a colloquium sponsored by the Phoenix LTER (Long-Term Ecological Research) presented by Steward Pickett of the Institute of Ecosystems Studies in Millbrook, New York. Pickett lamented the fact that ecology, traditionally an integrated discipline, had become too fragmented in the last 20 years. Ecologists are now pursuing a variety of research questions at different scales, focused on different processes and organisms, and conducted in different places, but lack the tools to integrate results into meaningful generalizations. Pickett argues that ecology should adopt a "culture of synthesis" including training in different ways of thinking, looking for analogies in areas outside one's expertise, re-thinking long-held assumptions, and reorienting traditional networks of interaction. Synthesis doesn't just happen. You have to work at it in the highly fragmented science of ecology. I walked away that day thinking I could have substituted geography for ecology and come to essentially the same conclusions.

Your correspondence convinced me that the problems of fragmentation and poor communication extend well beyond the boundaries of geography and that we can look to our colleagues in other fields for solutions. We can make a commitment to communicate more effectively with our colleagues inside and outside of geography, appreciate and articulate the disciplinary and societal context for the work that we do, and open ourselves more often to a culture of synthesis.

WILL GRAF: *Why Physical Geographers Whine So Much*
August 1998

It's a great day for geography. Physical geography in particular has enjoyed a remarkable renaissance over the past two decades, with renewed recognition and enthusiasm. Courses from introductory to graduate levels in physical geography are filled, opportunities for physical geographers abound in the world of government and industry, and the academic job picture is strong for geographers with an interest in the natural environment. Why then, several of my human geography colleagues ask, do physical geographers whine so much? The occasional undifferentiated angst among physical geographers lies in historical neglect, institutional support problems that lead to what seems to be unfair competition, connections to other disciplines, publication issues, and the "ism thing."

During the 1960s, when geography surged into the scientific era, there were relatively few physical geographers in the discipline. Even when their numbers and prominence increased in the 1970s, they were still a minority accounting for only a third of the AAG membership, and their connections with the rest of the discipline were tenuous. Human geographers, busy in their search for social and cultural explanations, saw their emerging physical colleagues as odd afterthoughts with unrelated research and teaching agendas.

Physical geographers also saw their research and teaching issues as distinct from those of the rest of the discipline, and the idea of over-arching geographic themes was lost. Physical geography's contributions were often inadvertently overlooked. This situation no longer exists, but the legacy is difficult to exorcise.

The late arrival of natural scientists in significant numbers also had unfortunate consequences in terms of institutional support in colleges and universities. For administrative purposes, geography had become a social science, and physical geographers, with their extensive field and laboratory activities rarely received the same level of material support that other geomorphologists received from geology departments, climatologists received from geophysics departments, or ecologists received from biology departments. Often, specialists in these other natural science programs teach fewer courses and receive more incentives for research than their counterparts in geography departments. These arrangements put physical specialists in geography departments at a competitive disadvantage and contribute to their discomfort in their home discipline.

Physical geographers are keenly aware of these differences, because they tend to be highly active in disciplines other than geography. Geomorphologists, for example, are often as active in the Geological Society of America as they are in the AAG: they attend and contribute to geological meetings, and they publish frequently in geological journals. The connections with other disciplines are satisfying and stimulating, but they also have the effect of further reducing the time and interest physical geographers have available for their own discipline.

The propensity of physical geographers to publish in the journals of related fields has meant that fewer of their manuscripts are available for strictly geographic publications. Because physical geographic specialists see themselves competing in a sphere wider than geography, they often put their best work in journals likely to influence specialists outside the field. As a result, their participation in geographic journals is not commensurate with their numbers. This behavior has led to a negative feedback mechanism in which physical geographers are less likely to submit to geographic journals where other physical geographers do publish. Editors have had little success in reversing the process.

The integration of physical geographers into the discipline has also been difficult because of the "*ism* thing." While their human geographer colleagues have been engaged in an ongoing debate driven first by Marxism, and then more recently by post-structuralism, post-modernism, and a host of other *isms*, physical geographers are perplexed, and not sure what all the fuss is about. Most physical geographers practice a form of neo-positivism, wary of the method's shortcomings, but also satisfied with the returns of their investment in the "scientific method." It is the method used by their competitors from other disciplines. They do not perceive a need to develop a post-modern climatology, for example, and they suspect (correctly or not is not the point here) that some *isms* are fundamentally anti-scientific. They feel a true disconnect when they contemplate their traditional scientific articles juxtaposed in a geographic journal with a post-modern treatise. This discomfort leads them to retreat to their usual position—outside it all.

It is important that the geographic family share at least some common themes, with physical and human geographers contributing to a better world through improved geographic knowledge. To achieve this end, and to address the issues outlined above, we need to take a number of actions, mostly at the individual level. Human geographers need to continue to work to make physical geographers feel welcome in their departments, activities, and organizations. Physical geographers need to stop playing the role of outsider and participate fully, especially in the AAG. They can do this the old fashioned way, one organizational step at a time. They must be active in their specialty groups, then volunteer for committees, then run for election as councilor or officer. They should nominate themselves and their colleagues, and take a genuine interest in the affairs of their discipline. If physical geographers question the wisdom of this approach, they should recall that their degrees, departments, and professional identities are geographic. There are no departments of geomorphology, climatology, or biogeography in the United States.

Physical geographers are not likely to have teaching loads different from their geographic colleagues, but they should continue to make the case that their work requires field and laboratory support, a case easier to make given the similar intensive technological needs of GIS that also must be recognized. Physical geographers ought to continue their strong connections outside geography, but not at the expense of their connections within the discipline. The richness of geography is in its diversity, and it would serve us all well if both human and physical geographers recognized this strength. If physical geographers want to see more of their articles in the discipline's journals, they will have to submit them, and suffer occasionally when they are declined. The empirical evidence (physical geographers like that sort of thing) strongly indicates that the types of articles published directly reflects the types that are submitted.

In summary, human geographers need to continue to invite their physical geographer colleagues back home, and to make affirmative efforts to involve them in the discipline. As for physical geographers, it is time to stop whining and get to work.

WILL GRAF: *Geographers' Too Small World*
October 1998

IT'S A GREAT DAY FOR GEOGRAPHY. More than ever before, geographic perspectives, technology, and knowledge are in demand outside our discipline. Consulting companies with a distinctive geographic slant are successfully addressing the issues in business, the environment, and social issues. Geographers in all levels of government, especially the local and state level, are practicing their professions and making meaningful contributions. The world outside geography is asking for our contributions. The question for those of us who are academic specialists is this: is our world view large enough to respond? I think we are too narrowly defined, and that to be successful contributors to solving societal problems we must take a more expansive view.

Academic narrowness stems from historical and institutional roots. The ever-finer divisions of knowledge and its teaching, along with an explosion in information and publication, has forced most academics to carve out small slices of the discipline, and then to focus only on that specialty. Early in this century most professors taught physical and human geography, often integrated in regional courses. By about 1960 we saw an increasing specialization whereby professors could teach either physical or human geography, but not both. Now we are to the point where many physical geographers cannot teach the entire introductory physical course because their specialized knowledge defines them as geomorphologists, climatologists, or biogeographers. During the period of the Cold War when the United States competed with the Soviet Union in science as well as in other arenas, we emphasized a reductionist approach to knowledge that piled up the publications, honed down the dominant paradigms, and emphasized narrowly focused teaching and research. With the advent of widely recognized society–environment issues and the end of the Cold War has come a new era in which society's most pressing problems are wide spread, integrative, and multi-disciplinary.

Geography is an obvious contributor to the solution of this newly recognized set of problems, ranging from democratization and global climate change to more local integrative problems such as pollution and economic development. I do not believe, however, that geographers are rising to meet these challenges. Instead, too many of us still write articles for ourselves, or the twenty other geographers in our specialty. We train our students to see and understand the details, but to neglect the context and the larger system within which these details play themselves out. Our vision often extends to the doors of our departments, but no further.

There are some steps we can take as individual student and faculty to participate in a broader geography. Autumn is a particularly important time to think about this issue, because it is now that many graduate students begin formulating their thesis and dissertation topics, and faculty members often do the same for grant applications. Often this formulation occurs by posing a question. We need to ask the right questions, questions that will generate answers useful to us as geographers, but also useful to other natural and social scientists, humanists, decision makers, and managers. This does not mean that we should abandon those things that are the core of our discipline: ideas about place and space, location, regions, scale, and the society–nature interface. It does mean that we need to generate questions susceptible to geographic inquiry that will produce answers needed by others outside the narrow confines of our discipline.

In short, new research questions, whether they are posed by students or faculty, should be able to pass the "in-law test" (those without in-laws may substitute a suitable alternative person). The test works like this: the next time you have dinner with your mother or father-in-law, and that person asks what you are doing these days, respond by explaining your presently active research question in no more than two sentences. If you can't explain it succinctly and clearly, or the response is "Oh, that's nice, won't you have another piece of pie?" you know you are in trouble. On the other hand, if the response is "Oh, that's interesting! I think that's important!" you know you are on the right track.

Generating research questions that pass the "in-law test" is not always easy because of several issues we must resolve. First is the tension between what we want to do and what is needed. Since most of us were trained to define problems narrowly, it is difficult for us to see their larger relevance. Sometimes narrow research and teaching are appropriate, but if their connections to someone who can benefit is obscure, it is hard to justify the investment of scarce intellectual capital. One way to determine the important teaching and research needs is to contact former students who now pursue their geographic careers in the real world, as practitioners. These geographers in business, consulting, and government have important insights into topics of immediate concern, and they know where the gaps in knowledge are.

The issue is the limitation that sometimes becomes apparent when we compare what we can do and what needs to be done. Most of us have knowledge and skills that can be valuable to those outside geography, but a broad application of what we have to offer usually requires us to learn new skills and to become involved with large groups. Many of us who were trained before 1990, for example, lack GIS skills. To be effective, many of us in that position need further training now, in order to employ our knowledge and experience using new technologies. Most professionals in other disciplines expect geographers to be familiar with GIS; there seems to be no good reason to disappoint them. Geographers must also learn to operate as members and occasionally as leaders of large teams. We can still work as individual artisan scientists, but once in a while we need to participate in integrative teams that tackle large scale issues in teaching and research.

The third issue is the relationship between the quality and the usefulness of our work. Research and teaching that are well done and that produce good products are not enough: those products must also be useful. The objectives of quality and utility are not mutually exclusive; rather they are mutually supportive goals. We can make geographic knowledge and a geographic perspective exceptionally useful, for example, in decision making and public policy arenas where we can contribute at may levels. Working in the policy area means, however, that we must accept a way of doing business that is very different from the academic world. In policy, our contributions are often not formally recognized; we must engage in negotiation more than teaching, and competition as much as cooperation. The so-called art of politics, something we try desperately to avoid in the academic community, is always part of the process. It is not likely that a single geographer will solve any of the world's great policy problems, but it is highly likely that each of us can contribute a small step toward solutions. Geographers can make a difference, and we should, but to succeed we must see a larger world.

REGINALD G. GOLLEDGE: *Moving on Up to "The Big Time?"*
May 2000

Geography is a small discipline. With only about 6,500 members of our professional association, we can hardly expect to be a leader in the sciences, humanities, or technologies. But we have made significant contributions in each of these areas. And we've done it with "small time" resources. Perhaps it is time to move up to "the big time."

Moving to the big time involves dramatically enhancing our objectives, changing the scope of our visions, and aiming at getting more resources. Easier said than done, but certainly feasible. So, how can we facilitate this transition?

Enhancing Objectives

We should no longer be content to be dismissed as "geography—the study of city names and sizes, capitals, and products." The last fifty years have been spent in going beyond that woeful image, but few people seem to know or care. "Geography? I hated that in school" is an all too frequent reaction when talking to people about our discipline. But the geography they hated is rarely being practiced today. Despite attempts to inform people about geography as it is now, geography's image is not positive. *Rediscov-*

ering Geography was a great idea, and while it has had an impact with the National Research Council and some government agencies, it hasn't taken the public by storm. It probably needs a popular version presented in an example-laden format and a storybook style to get the message to the general public. Or we need a traveling museum display, or some more innovative geography games (don't forget the effect of Carmen San Diego and Sim City, regardless of whether or not we academically approved them).

Changing the Scope of our Visions

Even while geographers engaged in fierce battles for domination of the field by supporting one approach or philosophy after another (e.g., determinism, positivism, regionalism, spatial analysis, social theory, post-modernism, information technology), we have maintained the usually prickly barrier of us versus the outside world. We claim to be the integrative discipline par excellence, but there's a lot of vapor here. Integrative science may be what geography is theoretically about, but, practically, segregated topical specialties dominate the discipline. If we are to change the scope of our vision, perhaps this is the first obstacle to address. How do we do this? Some have suggested disbanding the AAG Specialty Group structure and replacing it with more broadly defined interest areas—e.g., sustainability and vulnerability studies; global issues and local places; representation and analysis. Perhaps even the new *Annals* subdivisions (Nature and Society; Environmental Science; People Place, Region; and Models, Methods and GIScience) could be a starting point. If we are to think more in terms of being the natural link between human and natural environments, then having the discipline split into fifty barely communicating specialty groups may not be the way to the future. Certainly, if we wish to compete in the big time, we probably would have a better chance with larger and more coordinated interest groups than we have now. Deciding if this is feasible could be a project of an AAG task force. As of this time, our specialty group names don't coincide with any big time national movements—from digital earth to sustainable cities. Perhaps changing the way we look at ourselves will change the way others look at us and the way we envision the role of geography in the future.

Getting More Resources

The AAG, like other voluntary professional organizations, runs on a shoestring budget. The Geography and Regional Science Program budget at the National Science Foundation is ridiculously small for a discipline that includes extensive fieldwork and data collection in natural, built, and cognitive environments. When we think of being funded, most of us are happy to get $50,000-$60,000. In the larger scheme of science, that's peanuts. Not that we turn down such welcome funding, but it almost always means bare bones support and an inability to pursue grander goals. Geographers need additional (major) funding sources if we are to pursue the goals and objectives of a 21st-century human and natural science. In the NCGIA, the discipline has an outstanding model of what can be done with big time funding. Geography needs about ten more of these. Why only ten? Let's start somewhere and go for it! Prepare and act now. In 2002 there should be a mega-million initiative hosted by NSF to fund Infrastructure and Information Technology for the 21st century. Geography should not be satisfied with competing for one funded project—let's start planning now for a bunch of proposals! In addition, geographers need to start lobbying other government funding agencies and private philanthropic foundations. An occasional hit with Ford or Rockefeller isn't enough. We have quality personnel, we have societally and environmentally relevant and critical problems—let's market our interests, skills, and abilities to the big time.

A successful move to the big time can only help us achieve other goals. We'll never get geography reinstated in the major private schools as long as we are considered small time. We have to show proof of concept—show that we can contribute to significant problems and successfully compete for big funding—before we'll be able to change our small time image.

Obviously, change takes time. But the changes I'm suggesting here can be achieved in a short time. Most of it involves a change of attitude. We can start as a profession and as individuals now. Just review and enhance your objectives, enlarge your vision, and actively seek new resources. This has to be a grass roots movement. The AAG can help guide and offer advice. But geography won't move to the big time unless the bulk of the individual professionals that makes up the association want that change. What do you want?

SUSAN CUTTER: *Why Didn't Geographers Map the Human Genome?*
September 2000

AS FAR AS I KNOW, there is no central place theory to locate the amino acids that make a human genome. On the other hand, the discipline does have experience with mapping and visualization, so why haven't we been more involved in large-scale and highly visible research efforts such as the Human Genome Project? Since so many of us identify ourselves as geographers and something else—geography and hazards (hazards geographer), geography and politics (political geographer), geography and climate (climatologist), or geography and information science (GIS), this question becomes very relevant.

Many of us work at the margins of geography, practicing our craft in consultation with cognate disciplines. These multidisciplinary endeavors bring a diverse set of interests, skills, and methodologies to bear on a particular set of problems such as global environmental change, the widening gap between rich and poor, the global spread of AIDS, transformations of economic and political power, regional conflicts, or the spatial representation of phenomena. Interdisciplinary research, on the other hand, has the potential to produce new knowledge and move the scientific enterprise forward in revolutionary ways, such as the Human Genome Project. Interdisciplinary research requires us to think in different ways and to integrate knowledge among distinct disciplines.

I believe geography is the force that propels both multi- and interdisciplinary research. Geography is perhaps the only integrative science, and our knowledge and skills are sought now more than ever. But this wasn't always the case. Almost thirty years ago, geography was in a position to lead a new interdisciplinary effort called environmental studies. With one of our core traditions focused on nature-society interactions, it seemed like a natural fit and a perfect role for the discipline. Unfortunately, we failed to become the intellectual driving force behind environmental studies as new centers, institutes, and programs were being established in colleges and universities throughout the nation in the 1970s. Unfortunately, many geography programs were weakened on campuses in favor of environmental studies. This was not true everywhere, but for the most part, geography missed the environmental studies bandwagon.

In the 1990s another opportunity presented itself under the guise of global change. Funding agencies used the term "integrated assessment" as the nom de plume for identifying and evaluating the potential impacts of climate variability on specific regions. Why didn't they just say "regional synthesis," "geographical analysis," or "geography?"

Why did it take so long for geography to demonstrate its intellectual leadership on these issues? First, for too long the discipline has looked inward rather than outward. We tend to publish most of our scholarship in geographical journals (which is important for tenure and promotion decisions), but many of these outlets are not widely read outside the discipline. So, while we might be dong great stuff, few non-geographers know about it. The reverse is also true—if you publish exclusively outside the discipline, few of your geographical colleagues know your contributions. A balance must be struck, but at a minimum, we need to publish more widely, and equally important, read more widely within and between our subfields!

Second, as a group of scholars, we have marketed our distinctive skills poorly. For example, what other discipline save anthropology covers physical science, social science, and humanistic perspectives under one umbrella the way we do? Our ability to synthesize information from a diverse set of sources and apply it to the understanding of change in patterns and processes from local to global scales is unique, but we rarely tout this as a marketable skill or a contribution to research teams. We also possess good spatial information skills (geographical information science, remote sensing, visualization) that are in great demand thanks to many of you in the GIS community. However, in many quarters employees and researchers forget that the G in GIS stands for geographic information science.

Third, I don't think we were quite ready for "prime time" thirty years ago. For example, our students were often narrowly trained (in subject matter and methods). Departments were too rigid in their peer evaluations of scientific contributions and appropriate venues for publication. We didn't necessarily work on important public policy problems nor have much to contribute to national debates on contemporary issues such as the regionalization of poverty or the globalization of the informational industry. Finally, we spent a lot of time lamenting the fact that students and the public were geographically illiterate rather than doing something about it. We have made good progress over the last three decades in most of these realms, especially in the area of geographic education and public policy research.

Fourth, the discipline suffers from theoretical and methodological myopia. Instead of embracing diversity in our methods (quantitative, qualitative, analytical, descriptive) we often deride colleagues and devalue their scholarship if their methods vary from ours. The same holds for our theoretical approaches, be they based in social theories, spatial theories, or those aimed at understanding nature-society interactions. We need to be more accepting of different viewpoints, approaches, methods, and theoretical orientations ourselves if we hope to convince others of our pre-eminent role as an integrative science.

So what is stopping us from leading the charge today? There are many reasons, but I'll only mention a few. The first is the theoretical and methodological provincialism, mentioned earlier. Second, there are local, national, and international institutional structures that create insurmountable barriers for doing multi- and interdisciplinary research. In many institutions, geographers are dissuaded from engaging in such collaborative efforts because of political turf battles, resource allocations, and in some instances fear from other researchers that we'll take over because we have so much to offer! Third, there's the time issue. It takes considerable time and energy to engage in inter- and multi-disciplinary research, to garner resources to support it, and to promote geography as an integrative science to administrators and funding agencies. Those time demands present huge constraints for younger members of the profession still climbing the academic, corporate, or governmental ladders, and trying to balance professional duties with personal responsibilities. Finally, I think we lack a collective spirit of adventure, it is hard to retool and set off into intellectual terra incognita. We have the necessary tool kit and scholarly capital to sell geography as an integrative science. We just need to go do it!

M. DUANE NELLIS: *Millennials, Geography, and Future Worlds*
October 2002

IN TODAY'S DATA-RICH INFOSPHERE, geographers must prepare themselves for a new generation of students. We also need to recognize how the infosphere is altering the world around us. For baby boomers, broadcast TV, 78s, LPs, 8-track tapes, vacuum tubes, and mainframes were part of the electronics environment of their youth, whereas the students entering universities today have grown up with satellite TV offering hundreds of channels, MP3s, DVDs, microchips, personal computers, and the Internet. Numerous books and articles have been written on this next generation including Neil Howe and William Strauss's *Millennials Rising* (Vintage Books 2000), which describes changes in values and expectations this new generation may create.

I am reminded every day of these dramatic changes and the impact they've had on the millennial generation by my two sons. If we are to effectively engage this new generation, we need to develop more effective teaching strategies and cultivate new attitudes about information and technology. And, we need to understand the ways students access and utilize a broad range of geographic information and interact with others on a global scale.

This is not to say that traditional approaches to teaching geography are no longer relevant, but that we must be open to the opportunities digital media provide for interactive learning that effectively engages students and enhances learning. We must also be prepared to capture student interest through information-rich Web sites that stimulate students to think geographically, as well as recognize the breadth and depth of our discipline.

Geographers have begun to develop stimulating Web-based courses that are important complements to more traditional approaches, and that are likely to appeal to a broader range of students interested in contemporary geography. Ken Foote of the University of Colorado, for example, has developed a two semester, eight credit experimental course (funded through grants from the National Science Foundation) called "The Geographers' Craft." Central to this course is a hypermedia online electronic textbook. The course introduces students to image processing and GIS as tools used in geographic research and problem solving.

Many millennials are also involved in virtual worlds—digitally created worlds are the focus of numerous video and computer games played over the Internet by people throughout the world. Geographies of these places are being written and discussed, but the impact of these virtual worlds on our notions of space and place are just beginning to be understood. In some instances, geographers are beginning to create virtual reality landscapes by integrating geographic information sciences and remote sensing to model the dynamics of various surfaces.

Another fascinating dimension of the digital revolution and the Internet is the way it has changed how we view the world. The digital revolution has, in effect, created new spaces. A new sub-discipline in geography that focuses on the study of virtual spaces of the digital world, or cyberspace, is gradually emerging. There are a number of books and articles, as well as Internet sites, which explore the geography of cyberspace. This is bound to intrigue millennials. Martin Dodge, a geographer at the Center for Advanced Spatial Analysis at University College London, has directed some of the research on this topic, and there are now a number of classes being offered in the United States and other countries on the geography of cyberspace.

Cybergeography encompasses a wide range of geographical phenomena, from studies of physical infrastructure, traffic flows, and the demographics of cyberspace communities, to the perception and visualization of these new digital spaces. Research that documents the worldwide infosphere fostered by the Internet will surely be of growing importance as the millennials further transform the ways we interact within geography.

Geography and geographers need to be prepared on a variety of fronts to address the changing attitudes and expectations of millennials, to accept and explore the new geographies that millennials and others will create (e.g., the geography of cyberspace), and to engage in the study of these new geographies. Geographers must also develop new approaches to teaching geography that reflect cultural and education shifts triggered by the development of the Internet and the digital revolution, and recognize that these new geographies will affect the ways geographic information is marketed and utilized.

M. Duane Nellis: *Geographic Information Technologies and Future Geography*
May 2003

IN ONE OF MY EARLIER PRESIDENTIAL COLUMNS I wrote about the new millennials, a new generation of students involved with the emergence of cybergeography and Internet geography, that are creating new geographies using information from a multitude of sources and across a variety of media. In this, my next to last column as AAG President, I would like to expand on this theme in the context of geographic information technologies (GIT), particularly as they relate to global positioning systems (GPS), geographic information science (GIScience), and remote sensing.

These three areas of GIT have been of particular interest to me in my own research, specifically addressing the uses of GIT for analyzing questions associated with rural areas. From my view, these rapidly evolving approaches for spatial analysis are of growing importance to an increasing number of geographers in our discipline. As many have pointed out, GITs are more than just tools or techniques, they have matured to the point where they have their own body of knowledge and theoretical constructs that make them a science in their own right, especially as they related to remote sensing and GIScience. That is not to say they don't have value as technical spatial approaches to analysis of geographic areas, or that they don't still have geography and our approaches to problem solving at their core.

A general survey of research published in a broad range of journals shows many researchers utilizing various dimensions of geographic information technologies as part of their methodology in such geography specialty areas as political geography, economic geography and regional modeling, soils geography, geomorphology, water resources analysis, urban geography, and agricultural and rural geography.

At the same time, GIT advances are occurring at record pace as witnessed in various dimensions of our daily lives. The invention of real-time, interactive, and mobile GPS/GIS technologies, for example, has created new real-time geographic analysis and real-time geography. Such developments have led to significant advances in the ways spatial information is collected, mapped, and used within the discipline of geography. They are now at the heart of a vast array of real-time interactive mobile computing, geolocation applications and asset management, and wireless geographic services. As Doug Richardson wrote in a chapter of the recent book *Geographical Dimensions of Terrorism* (2003, p. 118),

> *"the integration of data intensive, dynamic geographic location, and timing technologies (e.g., GPS) on a real time and interactive basis with the previously static worlds of traditional GIS, has moved the science forward."*

The dynamic modeling and management capabilities of these new interactive systems have created far-reaching opportunities for geographic sciences from academia to government to business geography.

Within my own interests linking remote sensing and GIScience, there are new sensors, new applications, new challenges in data integration, and new outreach opportunities. In new sensors alone, there have been substantial increases in the resolution of commercial satellites. Satellites such as IKONOS, which produce high quality digital data, challenge us to ask new questions about our ability to map and analyze geographic areas, and to integrate associated findings (at much higher data volumes) into geographic information systems with new abilities to model landscape dynamics. In addition, ease of access to image data and processing worldwide is making great opportunities for extending our analysis to new areas.

Yet some see geographers using these new developments as "techies" who are missing some "pure" geography based on more traditional approaches to geographic analysis. In my opinion, nothing could be further from the truth. Those embracing GIT, and through this developing new science linked to geographic information systems, GPS, and remote sensing, have helped us move spatial sciences forward in new and exciting ways. Such GIT efforts have also allowed us to address the geographic problems of academia, business, and government, and have given some of our academic departments new and dynamic visibility on campus. Some of these academic efforts have focused on applications of GIScience, or more broadly GIT, to serve their university community, and in some cases their region or state. We at West Virginia University, for example, have been able to leverage our expertise in GIScience to create a State GIS Technical Center, and through this, enhance the base funding for our efforts to provide important spatial data services to state agencies. This has further stimulated investments by the university to strengthen the broad range of GIScience within geography, and allowed the department to develop certain strengths in community-based GIS, the use of GIScience in virtual worlds, and the use of GIT to understand the dynamics of change in rural environments. There are many similar opportunities for geography units nationally to seize opportunities associated with their leading role in information processing through GIT, and complement the mission of their college, university, or business, as it relates to use of spatial information.

Geographers must embrace future opportunities in spatial information sciences as it relates to GIT or others will fill this void in ways that may minimize the spatial science and analysis core of these approaches, and negatively impact what some view as the traditional base of our discipline and our relative position within the university community. As we look to the future, I believe we must: (1) take advantage of pervasive computing to further our discipline and analytical approaches, (2) play a more active role in information management at all levels and related geographic policy, and (3) foster interdisciplinary research using GIT that keeps geography and spatial science at its core. Through such efforts, I believe GIT, and indeed geography will only increase in importance in all dimensions of our future world.

❀ ❀ ❀

Chapter 10

GEOGRAPHY AND PUBLIC POLICY

❋ ❋ ❋

A consistent theme in columns during the 1990s was the need for geography and geographers to speak to the important issues of the time. The kernel of ideas that ultimately resulted in *Rediscovering Geography* (a National Research Council/National Academy of Sciences report), and *Global Change in Local Places* (Cambridge University Press) were first presented in these columns, which are reprinted here.

❋ ❋ ❋

SAUL B. COHEN: *A Time for Geography to Speak Out on Important Issues*
January 1990

ONE OF THE MARKS of the value of a scholarly discipline is the extent to which is has something to say about matters of societal significance. These are times of international upheaval and domestic drift. The debate has been joined over the consequences of the impact of the sudden end of the Cold War and the crumbling of East European communism, of the preparations for the Europe of 1992, of Japan's political stirrings in response to its economic preeminence, of the internationalization of the drug trade and environmental deterioration, and of the lack of economic progress in most of the developing world. For the United States, the next few years will present challenges that public policy cannot avoid. Will substantial declines in defense spending produce a peace dividend to be expended on needed new economic and social initiatives, both at home and abroad, or will it be used as an alternative to tax increases to reduce federal deficits and balance budgets? In many of our cities crime, drugs, illiteracy, and homelessness are located side-by-side with unprecedented excesses of wealth—the wealth of the nation and the world. In the face of such stark contradictions, are we facing a new wave of social and political tension, incomparably more wrenching than the urban upheavals of the 1960s? Environmental degradation goes hand-in-hand with consumer excess. How can we change a public policy that defers for tomorrow solutions to today's problems, let alone begins to weigh solutions for tomorrow's problems?

In some areas geographers are speaking out and are being heard— global warming and climatic change being the most prominent examples. But in general, in the United States the voices of geographers who have something to contribute to the national debate on policy are faint.

We are not alone with regard to the limited impact of our scholarly discourse. Sociology, anthropology, psychology, and history have influential figures whose theories and pronouncements contribute to the public debate. Nevertheless, they also lack the kinds of organized policy forums that permit systematic airing of views and reassessment of theories and strategies from their disciplines' vantage points. If government does not look to many disciplines for advice in the same formal fashion that it does from a National Science Board, or a Council of Economic Advisors, this does not mean the only recourse is hand wringing, blame casting or letters to the editor. As geographers, we should be addressing the challenge of how best to marshal our intellectual resources to bring our insights to the public's attention.

As a start, our most important assemblage—the Annual Meeting— should become a more prominent focus for mobilizing individual and team efforts around overriding problems. The two Presidential Plenary Sessions that will be presented at the Toronto meeting—"Sharing the North American Continent" (chaired and organized by Kenneth Hare) and "Sharing the Global Village" (chaired and organized by Gilbert F. White)—are more than simply mass sessions. They are a call to action—a challenge to organize our research efforts in greater concert around problems with which society is grappling. It is encouraging that each plenary has stimulated approximately 40 sessions following up on specialized aspects of the general themes. From amongst those many papers, we should be able to identify important directions to pursue in larger-scale fashion.

But a forum at a professional meeting has a limited life in its impact upon the public consciousness, and even upon the profession. We need mechanisms that will promote sustained research into policy issues. If by the year 2000, 75 percent of the US population will be concentrated within 50 miles of our coasts, what do geographers have to say about the spatial consequences of rising sea levels? If the moratorium on construction of nuclear power plants leaves certain parts of our country with energy deficits, what are the spatial implications for the long-term utilization of human and natural resources? If foreign capital overwhelmingly piles up in a handful of cities and largely flows into the real estate sector of the economy, what will these land use distortions mean regionally and nationally? If humanity's impact upon coastal erosion distorts natural equilibrium forces, do we have barometers that are sensitive measures of the rhythm of landscape change responses to economic and natural forces? We could go on and on with questions, but the point has been made.

In the summer of 1988, Gilbert White and I chatted at the IGC in Australia about how our Association could be more effective in tackling society's overriding problems—in effect, how to put more meaning into the phrase "geography in the service of the nation and humanity." Since I saw this year as a year of organizational restructuring, my sense of priorities may, therefore, have been inadequate. Fortunately, we have now addressed most of our organizational issues and can turn more fully to programmatic initiatives.

More recently, I met Gilbert again at the Binghamton Applied Geography Meetings. He told me that several years ago he had suggested that the AAG consider developing an annual Geographical Report on the State of the Nation. He and Tom Wilbanks sought to elicit some interest in the Report, but without success. I do feel that this is a worthy focus for American geography, and that the timing is right to give this priority within the Association.

Some of you doubtlessly have had some thoughts along these lines—Ron Abler and I would welcome hearing from you. For those who indicated an interest, we might hold an informal idea-exchange session at Toronto. In any case, there is need for geography to find a systematic public policy forum for pertinent theories and insights. If we develop such a forum, we can begin to address the challenge of bringing the nation to a higher level of geographical understanding and sophistication.

THOMAS J. WILBANKS
December 1992

The Honorable Bill Clinton
President-Elect of the United States of America
Governor's Mansion
Little Rock, AR 72201

Dear Mr. President-Elect:

As President of the Association of American Geographers, I congratulate you on your election and offer you my personal commitment, on behalf of the nation's professional geographers, to work with your new administration toward solutions to the problems facing our country. Geographers have a great deal to contribute to your efforts in economic revitalization, education reform, environmental management, and other important policy arenas; and we hope your team will call upon us to help.

A starting point, of course, is geography's place in the National Education Goals (which you personally helped to shape) as one of five core subjects for America's schools. In support of these Goals, all of geography's national organizations are working together to define "world class" national standards, to develop frameworks for testing what students know and can do, and to help state and local school systems implement this educational reform in a positive, relevant manner. We believe that geography's role in the Goals will lead to significant payoffs in student readiness to be productive workers, well-informed citizens, and happy, fulfilled people throughout their lives.

More than this, however, we geographers believe that we know some things that can help you and your national policymaking. As examples, let me suggest a few simple truths, based on solid research by geographers and others, that are indicators of national needs where we might be able to contribute.

1. *National economic recovery depends on increasing our viability in the global economy. Our economic problems as a nation reflect the fact that, while many other industrialized and industrializing countries have known for generations that their economic well-being depends on resources and markets beyond their own borders, we have tended to focus on ourselves during our long age*

of abundance. Now, we are realizing that our jobs and incomes are directly related to our ability to compete with workers and firms in the international marketplace. Meanwhile, we see our economy in this marketplace taking on many of the characteristics that we have associated with developing countries: exporting raw materials and agricultural products, importing capital and new technologies. Turning this situation around requires, as a minimum, that we get much better informed about the rest of the world: market conditions and consumer preferences, comparative advantages both here and there, reasons why we are sometimes not competitive. In part, we are slipping because other countries are simply doing a better job of international market research. They know the rest of the world better than most of us do, and therefore they do better in foreign markets. More broadly, though, we are suffering because too many of our firms and fellow citizens simply do not understand how we are linked with other countries economically and how our financial prospects depend on responding to these linkages more positively and innovatively than our competitors do. Geography would like to help solve this problem, and we are ready to suggest some strategies for accelerating the process.

2.   *There are significant regional differences in the causes and effects of the current recession. As you know more than most, after your more than a year crossing America, the recession has a different meaning in Michigan than in Arkansas, a different meaning in California than in New York. Different people are hurting for different reasons, and policies targeted on some needs will miss others. A simple example is the minimum wage, which can be painfully high in Mississippi's economy while it is absurdly low at Alaska's cost of living. In our diverse country, effective public policy needs more sensitivity to regional differences and geographic balance than it has usually had. In fact, our Presidents have often found it easier to get geographically detailed information about varying policy implications between different foreign countries than between different US regions. Again, we would like a chance to help fix this constraint on sensitive executive-branch policy making in a governmental system where every vote in the legislative branch is cast by a representative of a geographic area.*

3. *Our long-term well-being depends directly on a harmonious relationship with our environment. You and Vice President-elect Gore certainly appreciate this point; but one of the biggest challenges of the next four years will be to set policy directions related to such global issues as disruptive environmental change and such national issues as waste disposal, in a new era of understanding that we must treat the environment as our partner rather than our servant. More than most other fields of learning, geography has a long tradition of viewing nature and society as inextricably linked, paying attention to both sides of the partnership and to their relationships—and special attention to the human dimensions of these relationships, in terms of both limits and opportunities. Our research has ranged from population-resource balances to environmental hazards and conflict resolution. In these kinds of connections, we believe we are in an underutilized resource for our nation.*

4. *There is no fundamental reason why basic human needs in this world cannot be met. We do not have hunger in this world because there is too little food. We do not have a lack of shelter because of a shortage of building materials. We do not have a lack of investment in needy areas because there is too little money in the global financial system. We do not have inadequate social services for fellow citizens because we cannot afford them. Time after time, we fail to meet human needs because our distribution systems are inadequate: poorly informed, slow to respond to imbalances between supply and demand, and lacking incentives to do much better. Some of these distribution systems are geographic; some are political-economic. We geographers know more about how to improve the geographic ones, of course; but our research has touched them all, and we want to join in a serious effort to improve distributional efficiency, equity, and justice wherever needs continue to exist.*

5. *Cultural diversity within US society should be a source of strength rather than of friction. Finally we agree with those who have suggested that the United States is not a "melting pot," nor should it want to be. How boring it would be if Mena, Arkansas, were the same as Marin, California, or if Ysleta, Texas, were the same as South Philadelphia. We Americans believe in equality of oppor-*

tunity for all, and we expect different groups to respect laws that preserve peace and order. But we should stand for more than a tolerance of cultural, ethnic, and religious differences; we should work toward an active appreciation of their value for our society, our economy, and our lives. Variety stimulates learning. It generates a wider range of ideas about solutions to problems. It offers more alternatives in a society that associates vigor, efficiency, and opportunities for fulfillment with the freedom to innovate. We need to unlock the potential that our diversity offers us and turn it into one of our national advantages. Once again, geography can help, because we have always taught—and believed—that diversity is the stuff of discovery, not of threat.

In these ways and many others, geography has a lot to offer national policymaking and problem-solving in the United States. Obviously, we geographers are a diverse group of people; but more than most other scholarly disciplines, we are rich in professionals with perspectives very much like yours. We care about places and believe they are important. We specialize in the study of linkages, both between our country and the rest of the world and between people and their environment. We are impatient with intellectual boundaries. We are oriented toward identifying and solving real-world problems—embracing their complexity rather than assuming it away.

In a spirit of cooperation and support, we will be pushing your administration to give us a bigger change to combine our knowledge as professionals with our concerns as citizens, working with you to improve the human condition in this country and throughout the world.

Bob Kates
May 1994

Last month I described the Association's agenda as a trinity of goals: achieving geographical literacy; making geographical and geographer-led interdisciplinary knowledge inform social purpose; and strengthening the discipline by strengthening our connections. This month, I want to address the second: the need to link the immense reservoirs of geographic knowledge of peoples, places, and processes with the continuing process of social decision and action. Let me begin with a bit of our collective past.

"On a global scale, wherever economic development is seriously considered, there is recognition of the world community's failure thus far to find effective ways of closing the gap between rich and poor nations. The twin problems of resource use and population density continue to worsen . . . At home, the tensions between racial groups, regional groups, and the intellectuals and hardhats become deeply divisive and strong. Our cities decay physically while spaghetti-like expressways carry the upwardly mobile to suburban havens. We are fast becoming disillusioned with GNP as an index of quality of life."

"It is no longer academic or fanciful to pose again and again the question of whether the world society . . . can survive . . . I would be delighted [if] twenty years from now we can all look back with amusement to those harassed days of 1970 when we entertained serious doubt that man could avoid a nuclear holocaust or genuinely prevent global disorganization or keep from fouling his nest irreparably."

Thus spoke Gilbert White in 1969 and published in the May 1972 *The Professional Geographer*. In that talk, he went on to advocate various ways geographers might direct their research toward possible social implications, how the Association might support efforts at relevance, and how the university as an institution might also need to change. He then concluded with this stirring challenge:

"Let it not be said that geographers have become so habituated to talking about the world that they are reluctant to make themselves a vital instrument for changing the world . . . What is important is where we stand in relation to the tasks of society. Little is to be gained by critically pointing fingers at white faces in textbooks, at vapid generalities about world power, or at observations about resources and man that are perfectly true, perfectly general, and perfectly useless . . . What shall it profit a profession if it fabricates a nifty discipline about the world while the world and the human spirit is degraded?"

Now, 25 years later, many of his observations still resonate, even in a post-Cold War world. But there is also some positive change. A sense of crisis still persists, but a critical geography that examines many of those crises has strongly emerged. And equally, if not more important, we have learned much that is relevant to addressing the crises of our time.

Our three great traditions that have focused on the human use of the earth, the connectiveness of space, and the distinctiveness of place provide us with much middle-range theory, time-proven methods of analysis, and an enormous body of observation and empirical study. Thus we can speak of the *Earth Transformed* with its 130-year history of understanding the human modification and transformation of the earth. We can think of geography as the science of flows, especially the flows of human life: people and their life-supporting flows of energy, materials, and information. And we have observed and elaborated on those distinctive qualities that govern the diversity in environment, habitation, livelihood, and meaning that make for a sense of place and how one place differs from another. Today, we geographers have a distinctive, albeit not unique, set of contributions to make to "the tasks of society."

Thus, the Association's efforts are to act on what we already know, to place our geographic knowledge in the public service and in behalf of the civil society, and to better link disparate sectors of our research effort. We have four initiatives underway:

The first originated in an effort of Tom Wilbanks to acquaint the incoming administration with the value of geographical expertise by addressing letters of welcome and offers of assistance to the newly designated Secretaries and Administrators of departments where geography had something to offer. A great idea, but we found it difficult to carry out because there had been little previous effort to order what we know in ways that it might be useful to those charged with doing something about it.

Thus to systematically address that issue we have turned to our 43 specialty groups, particularly those that bring together geographer interests in topics of high public concern such as energy, hazards, medicine, and regional development. We hope that the appropriate specialty groups will create public issue response teams that will try to match geographic research and expertise with public issues, and to encourage more relevant research and its public expression.

And for such relevant research, we will initiate our own project designed to link the great international effort to understand global environmental change with and understanding of other global economic and demographic changes and with the distinctive knowledge that geographers possess of local places. The Association proposes to examine how economic restructuring and population growth and decline contribute to land use changes and trace gas production in smaller places, and how sensitive such places are to global environmental change. To do so, the project will draw upon the long-term regional commitment of selected college and university teaching institutions, the underutilized talents of their faculty, and the reservoirs of local knowledge centered in their geography departments.

We also have two efforts underway to increase the demand for our knowledge. Before the year is out we hope that we can arrange a meeting between the Presidents of the AAG and NGS and the appropriate staff of the Office of Science and Technology Policy to increase our access to scientific advisory and policy boards. And a major report will be forthcoming from the National Research Council Committee reviewing the discipline, the Rediscovering Geography Committee. The report will be directed toward the science and users of science community and will be built around what geography can contribute to the great questions of both science and society.

Finally, it is important to acknowledge a tension in all these undertakings, between two sets of equally socially-concerned geographers, what I would designate as the fixers, and the critics. The difference was sharpened for me by an exchange between my predecessor and two critics, in which Tom Wilbanks said: "the central issues—for most of our fellow citizens as well as our policymakers—have to do with jobs and income . . . and we are not contributing very much to the discussion of these issues." He went on to advocate strengthening our links to business and industry and to research related to current economic needs (President's column, February 1993). To which, Jones and Roberts replied by asking: "Are we reading the same literature? Hasn't our discipline been a leader in charting and theorizing transformations in the space economy during this period? . . . Aren't we now engaged in serious reflection on the global economy, the restructuring of US industry, and the upheaval of old geographies that thereby ensue?" (April 1993)

There are clearly ideological differences between Wilbanks and Jones and Roberts, although perhaps not as large as any of the three might imagine. More important, there is a functional difference, one that many of us share, between seeing our major role in addressing social issues as fixers who can actively place their knowledge in direct efforts to make things better and critics who see little hope in making things better and certainly not until their basic critique is widely accepted.

Now this distinction is clearly an over-simplification, as in all roles many of us are both critics and fixers. But I would argue that recognizing the difference is a first step to appreciating each other more and the ways in which we can be helpful to each other. For indeed, the Association's agenda is to seek ways in which the best of our understanding, critical or otherwise, can address human survival.

WILL GRAF: *How Geographers Can Connect with Policy Makers*
November 1998

IT'S A GREAT DAY FOR GEOGRAPHY. Opportunities for geographic input to decision-making abound. In the United States we make collective decisions within large geographic aggregates of smaller spatial units, ranging in scope from metropolitan area organizations, airshed, and watershed councils, to interstate compacts for water and transportation. Almost every policy decision by such governmental organizations touches on our expertise, yet geographers are too often absent from the process. Many geographers avoid participating in public policy service because it is so different from the working environment to which they are accustomed. Policy work is often poorly focused, and the rules of the decision-making process are in constant flux. Unlike the academic world, where the faculty member occupies a clearly defined position of authority (such as it is), the expert in the public policy debate must constantly prove his or her mettle. Often, public involvement in policy processes means dealing with individuals or groups who are vehemently opposed to a possible solution, or even to the debate itself.

I have been fortunate to have had opportunities to mix my research and teaching with policy experiences, as an expert witness in legal proceedings, as a member or chair of several National Research Council and similar boards and committees, and recently as a member of a Presidential commission on river policy. I have learned the following lessons:

1. *You must approach them; they won't come looking for you.* Initially, you must put yourself forward to make yourself and your expertise known to decision makers. Opportunities for entry-level volunteer (read "unpaid") involvement include service on technical advisory panels, review boards, and work as a commentator on management plans by governmental agencies. Opportunities for paid service include consulting activities identified through professional contacts, including

former students. In many cases, the decision makers simply don't know you are available. Once they identify you, they often welcome your input.

2. *Identify opportunities in unconventional ways.* Search for opportunities for policy-related service in newsletters, Web sites of governmental agencies, non-governmental advocacy organizations, and general news outlets, including regional publications. If you do field work in a particular area, introduce yourself to resource analysts, planners, and managers who deal with the area. After you meet with them, send them copies of your publications or course outlines, and offer your services.

3. *Have something to offer.* Credentials give your advice legitimacy and enhance the potential influence of your contribution. Any citizen might adopt an advocacy position, but geographers can offer important additional input to decision making by applying their knowledge gained from research and teaching. Your vita is your calling card, and it must reflect the potential for your contribution.

4. *Make the most of your strengths.* By keeping your public policy work aligned with your teaching and research, you maximize your contribution and strengthen your expertise. Teaching, research, and public service should mutually enhance each other.

5. *Search for benefits to you.* Policy related efforts usually don't offer financial rewards, but they definitely offer benefits. They provide access to new data and ideas, contacts to develop into internships and employment for students, sources of funding for later development, and opportunities to extend professional contacts. In the long run, the career advantages offered by these benefits are worth more than short-term contract money.

6. *Be forceful and convincing.* Once you secure appointment as an advisor or committee member, make substantive contributions. This arena is not for shrinking violets. Don't shortchange the process: the people you serve with expect you to give your time, energy, and enthusiasm to the decision at hand. Figure out how the process works and who makes the important decisions. Learn the system.

7. *Ask questions.* Most policy processes are overloaded with acronyms, strange and wondrous short hand language that is everyday lexicon for the policy maker and mostly unintelligible to everyone else. What you don't know can limit your effectiveness. If you don't know what

FACA (Federal Advisory Committee Act) and FOIA (Freedom of Information Act) mean and how they work, ask. You will have plenty of opportunities to avoid problems for your policy colleagues by avoiding scientific jargon and speaking in plain English when addressing technical issues.

8. *Develop a network.* Use the service opportunity to develop and expand your network. Introduce yourself to others in the process, and get to know them while making sure they get to know you. Word of mouth is the most important method of advertising your contributions in the policy area. The more contacts, the better.

9. *Be prepared to lose.* In all policy and decision making processes there are short-term winners and losers. The positions you take will often not prevail. You should be prepared to compromise, and to lose. Positions that seem obviously correct to you from a technical standpoint may be culturally unacceptable, too expensive, or politically problematic.

10. *Win with grace.* The positions you take may sometimes be successful, but crowing about it creates resentment among those who disagree with you, and you may need their support on the next issue. They are your potential network, so rubbing it in is risky business. In many cases, you will not get credit for successes because the decision maker you advise will use the credit for his or her own political capital. Your reward is your influence.

Involvement in public policy issues benefits us as individual professionals. For geographers in business, the involvement develops valuable contacts with potential customers, and increases our visibility to those who might retain our services. Those of us in the academic world can import our policy involvement directly into the classroom, demonstrating to our students that the geographic principles we offer have direct social and environmental relevance. Policy experience is essential to the effective researcher because experience in dealing with decision makers indicates the truly important research problems. Participation has a moral imperative as well. Many professional geographers have spent their entire lives developing knowledge and experience, often supported by taxpayers, and we have an obligation to share the results of that investment with those who pay the bills. Geography is well worth that investment, and we can make a difference.

SUSAN CUTTER: *Taking Some of the Ivory out of the Ivory Tower*
March 2001

I WAS RECENTLY ALERTED to the January/February 2001 issue of *Foreign Policy* by a couple of our members. The cover of the magazine bears the image of smiling African children with the title "A Case of Bad Latitude: Why Geography Causes Poverty" and includes an essay by Ricardo Hausmann called "Prisoners of Geography." If the regional stereotype on the cover doesn't get your blood boiling, just wait until you read the article written by Hausmann, a professor of the practice of economic development at Harvard University's Kennedy School of Government!

Professor Hausmann has "rediscovered" economic geography without reading or referencing any contemporary work by geographers. Instead, he relies on antiquated notions of the discipline fostered by poor scholarship, the insularity of elite institutions, and intellectual arrogance fostered by economists who are touting themselves as the new economic geographers. His rendition of the demise of geography departments at Harvard (in a sidebar called "Locational Correctness") and elsewhere (Michigan, Northwestern, Chicago, Columbia) is just plain wrong.

The prominence of the article in a widely read journal illustrates two important issues. First, it is a superb example of why Harvard needs to reintroduce geography into its curriculum in order to inform faculty and students on the nature of geographic inquiry and the powerful tools and analytical techniques that guide and enrich understanding of the world's people and places. Second, the article highlights one of our failures—limited engagement with the broader community of scholars, policy makers, and business leaders.

While we can lament poor scholarship on the part of practitioners of other disciplines, what are we doing (individually and collectively) to reach colleagues in cognate fields or to influence public policies? How can we impress the intellectual and policy communities with our theoretical, analytical, and methodological advancements if we talk only to one another in highly specialized journals, abandoning the broader audience for some of our work? While many of our colleagues are recognized beyond the boundaries of our discipline, many more of us should be reaching out to advance geographic perspectives through our publications, teaching, and service. We have intellectual capital that has considerable worth, but we're just not marketing it as well as we could or should.

Let's examine some of our successes. Geography has been the driving force behind GIS and its newer reincarnation, GIScience. Data arrayed in geo-coded form ready for processing are now the norm, not the exception in governmental agencies from the local to the federal level. The standards for geocoding and classification of data promulgated by the Federal Geographic Data Committee were largely driven by and staffed from the geographic community. We have, as a discipline, influenced governmental policy on geographic data.

Another success story is educational policy. The grass-roots efforts of the National Geographic Society-sponsored State Geographical Alliances have helped transform and upgrade the achievement standards for the social sciences in grades K–12. Geography is no longer absent from K–12 curricula, nor entirely subsumed under history. There are established milestones of geographical knowledge that are required for advancement. There is now even an Advanced Placement course in human geography.

A third success comes from the environmental policy field. Recently, the AAG undertook an effort to place geography in a more prominent position in the public policy arena. Geographers have often played a leading role in natural hazards research and its application to public policy (see the January *AAG Newsletter* recognizing the contributions of Gilbert White). In an effort to elevate natural disaster reduction to a national priority and to influence public policies focused on natural hazard risk reduction in Congress, a Natural Hazards Caucus Work Group (*www.agiweb.org/workgroup*) was formed. The AAG was one of the 35 founding members of the caucus. Geography is now clearly identified as one of the players that can provide scientific expertise in support of these activities. We are in good company. The American Geophysical Union, the American Meteorological Society, the Seismological Society of America, and the University Corporation for Atmospheric Research, to name just a few, are also founding members.

Where are our failures? It appears to me that most of human geography has not made its voice heard within the policy community, or if it has, few people are aware of contemporary contributions. Where has the discipline influenced social policy? Urban policy? International relations? Health care? Economic policy? Reapportionment? Why is this? First, there is a mismatch between the topical interests of practicing human geographers and those of the policy community. Second, the preoccupation with theoretical and methodological issues within geography has widened the divide between academics and practitioners. Third, the outlet for human geographical scholarship remains cloistered within narrowly-framed journals and publishing houses. There is rarely any attempt to reach a broader, more policy-oriented audience.

We should make our voices heard and tout our abilities to contribute to contemporary public policy issues and important intellectual debates. This can be done through good scholarship placed in highly visible (often non-disciplinary) outlets, persistence, and the proverbial good timing. We need to become more aggressive and take advantage of opportunities to advance geographical inquiry whenever we can, whether it is in the halls of Congress, within mission agencies, or, especially, the Ivory Towers of Cambridge, MA.

ALEXANDER B. MURPHY: *Geography in an Uncertain World*
July/August 2003

THE OFT-QUOTED SAYING, "may you live in interesting times," is of questionable derivation (in all likelihood it cannot be attributed to Confucius), but its popular connotation captures some of the uncertainty that most of us feel at this time. From almost any vantage point, the global situation appears more volatile now than it did a few years ago. Closer to home, most of us are confronting the impacts of the recent economic downturn in the communities where we live and in the institutions where we work.

Against this backdrop, a natural instinct is to hunker down in hopes of weathering the storm. Adopting a defensive posture is sometimes important, but at the current juncture it is not necessarily the posture that will serve geography the best—or the larger society, for that matter. We are a discipline that is clearly on the rise. We are expanding in important ways, and other scholarly and professional communities increasingly regard us as a source of serious research, ideas, and insights. The time is ripe for us to make a compelling, non-defensive case for what geography has to offer.

Making such a case is clearly important, given the current funding crisis in higher education. Almost anyone teaching in a geography department in North America is concerned about the impact of budget cuts. In a recent presidential column, Duane Nellis outlined several things programs need to emphasize if they are to avoid debilitating cuts—expanding student credit hours, participating actively in campus affairs, devoting time to alumni relations, etc. These are all important, but even as we pursue them, we must also pay attention to the larger context of how geography itself is understood and viewed.

Any administrator who is not at the extreme "bean counter" end of the spectrum has an academic/institutional vision. Adopting a defensive posture is unlikely to appeal to that vision. Of course it is important to let people know about the good things that programs have done in the past and about the negative consequences of serious cuts. Yet such arguments are unlikely to win the day unless they are part of a larger vision of the value of geographical understanding and scholarship in the contemporary world. We need to demonstrate, in clear and incisive ways, geography's expanding influence and society's need for geography.

It is not difficult to draw attention to geography's expanding influence. Geographers are visible contributors to scholarly exchanges on topics ranging from long-term environmental change to globalization. Current debates in international relations are infused with geographical assumptions and ideas. Geographers have pioneered the development of a host of new geospatial approaches and technologies, even as geographical concepts have played a prominent role in the "cultural turn" in the social sciences.

Geography's institutional standing is also on the rise. Over the past decade the discipline has moved from a marginal to a central position in the National Academy of Sciences–National Research Council. Geography's position at the National Science Foundation is stronger than it has ever been. The College Board has added Geography to its Advanced Placement Program, and the names of geographers increasingly appear among the list of grantees of prestigious organizations and foundations. Even in these troubled economic times, there is news every few months of a new or expanded geography program at a college or university.

It is essential that administrators know about these things, for they paint a picture of a discipline that colleges or universities ignore at their peril. The case for geography cannot stop there, however; for ultimately it must rest on the intellectual and practical merits of geographical inquiry. To put it another way, our challenge is to convey the utter hopelessness of confronting the issues and concerns of our time without the benefit of geographic understanding and analysis.

In meeting that challenge, our understandable fear of being seen as a discipline focused on place-name memorization may make many of us hesitant to invoke popular conceptions of geographical ignorance. By avoiding this issue, however, we miss the opportunity to build on the commonsense notion that intelligent engagement with the world requires some understanding of how Earth's physical and human components are organized, how people live in and use particular parts of the planet's surface, and how places relate to one another. Of course, we also need to highlight how advanced geographical inquiry can shed critical light on the growing gap between rich and poor, the potential consequences of climate change, the impacts of humans of the environment, the nature and implications of ethnic conflict, and much, much, more.

All this brings me back to my opening reference to current global and local uncertainties. The communities, states, and countries in which we live are facing difficult and important decisions on matters ranging from the provision of social services to the development of appropriate responses to terrorist threats. As citizens and government officials confront these decisions, geographical understanding is critical. Wherever one may stand politically, it is clear that there cannot be a serious, meaningful discussion of the current situation in Iraq if (as was recently suggested) Iraq's internal character and regional situation is thought of as being analogous to Japan's after World War II. In a similar vein, a thoughtful or productive consideration of fire suppression in the American West cannot proceed without some understanding of the physical and human aspects of forest dynamics that come from geographical analysis.

The more that geography becomes part of the public debate over where our society has come from and where it is going, the more geography will be strengthened, as will society at large. Enhancing the discipline's position in public debate would thus seem to be a fitting goal as the AAG enters its second century. It is a cause to which we can all be committed, and it is at the forefront of my agenda as I begin my presidential year. I look forward to working with many of you in pursuit of this end.

ALEXANDER B. MURPHY: *Enhancing Geography's Role in Public Debate*
December 2003

EACH DECEMBER I face a moment of discouragement when I scan *The Economist's* annual list of influential books published within the past year. Rarely do my eyes fall on even a single title in geography—even though the list includes such categories as "Politics and Current Affairs" and "Science, Technology, and Philosophy." Geography still seems to fall below the radar screen of many influential publishers, journalists, and policy makers.

I raise this point not to set off another round of collective hand wringing. To the contrary, there is much to celebrate at the moment. Geography's profile has expanded substantially over the past two decades, built on everything from the GIS revolution to the spatial turn in neighboring disciplines. Moreover, the scholarly work of geographers is increasingly well known outside our field.

Yet for all of our successes, we still occupy a relatively minor role in the public conversation about society, politics, and the environment. *The New York Times* does not feel the need for regular commentary by a geographer, as it does an economist; geographers are scarce at the major think tanks and policy institutes inside the Beltway; and with the notable exception of Harm de Blij's stint on "Good Morning America," geographers are a rarity in the broadcast media.

What can we do about this? The most important response is the least flashy: we need to do excellent work on matters of importance. We can only enhance our stature by producing some of the best books and articles, by teaching some of the best classes, and by offering some of the best ways of tackling applied problems through the use of geographic techniques and perspectives.

Our numbers are small in comparison with many other disciplines, so if we care about the role of geography in public debate, it is also incumbent on us to be as strategic as possible in our efforts to reach a broad audience. Three things occur to me as particularly important in this regard: publishing books, connecting our teaching to the issues of the day, and writing opinion pieces for wide dissemination.

At a time when we are mourning the death of Terry Jordan, it is appropriate to remember a key theme of his AAG presidency: the importance of books. Much that happens within geography does not lend itself to book-length treatment, but when it does, it's hard to beat books for reaching a broad audience. Think of the impact people like Jared Diamond or E.O. Wilson have had. The general public knows them not for their articles, but for their books. Of course, geographers have produced some wonderful books, but the paucity of geography titles in *The Economist* reflects not just a lack of demand, but a limited supply as well. This is why I championed AAG book prizes when I was a National Councilor in the late 1990s, and why I think we need to continue to push ourselves to produce books.

As for our teaching, I would be the last one to propose a narrow "relevance" test for education. I have long been a proponent of a broad, liberal arts education, for an over-zealous concern with relevance can undermine the broader intellectual foundations that are necessary to create an educated citizenry. Yet there is nothing wrong with making connections between the things we study and the issues that are shaping the contemporary world. Given the appalling geographical illiteracy that is so evident these days in the public arena, making those connections is a matter of signal importance. When we teach about the cultural diversity of a place like Iran, we should not simply be telling our students that it is a place with a distinctive language and religion; we should be talking to them about what that means for simplistic understandings of "the Islamic World," or for the prospects for regional integration in the Middle East.

One of the most important things we can do is write about our work for a general audience from time to time. The editorial pages of newspapers are logical outlets. From the applied sphere to the theoretical, geographers are working on a myriad of interesting and relevant topics. What if, just once every three to five years, every professional geographer sat down and wrote an op-ed piece for the newspaper about the connection between their work and an issue of current interest? For such an endeavor to help raise geography's profile, the geographical dimension would have to be made explicit. But if done right, the cumulative effect could be quite striking.

So let me end this column with a challenge. If you have not produced an op-ed piece in the last few years, plan to write one. Soon. Identify yourself as a geographer and explain how geography is integral to assessing the problem you are raising. Of course it takes a special writing style (and often a certain amount of simplification) to do interesting, provocative pieces. To encourage this further, and to help geographers with the mechanics of the process, I would like to organize some sort of workshop or session at the Denver AAG meeting (Philadelphia is too full) to address this matter. Suggestions about what this might look like and how it might be carried forward are welcome.

❋ ❋ ❋

Chapter 11

NATIONAL VISIBILITY

AND THE PUBLIC ROLE OF GEOGRAPHY

❋ ❋ ❋

Not surprisingly, a recurring theme in columns is the need to engage a broader audience in order to improve the prestige of the discipline and the popular reach of geography. Many mechanisms were suggested to pursue these objectives as this group of selections illustrates.

❋ ❋ ❋

RONALD ABLER: *Of Levers and Geographers*
December 1985

"GIVE ME WHERE TO STAND," challenged Archimedes in his discourse on levers, "and I will move the earth." With a proper fulcrum and an appropriate lever, a small effort can be translated into a powerful or even irresistible force.

Geography is a small discipline. As noted in my October essay, the AAG is the smallest of the professional organizations with which we compete for resources and recognition, and one of the Association's priorities is to enlarge the geographical community.

Geography is also diverse. Some geographers study the world as social scientists, others as physical scientists, and others in humanistic terms. That diversity is an asset, but it further divides the small force geographers bring to bear in commerce, industry, government, and academe.

How can geographers exert leverage to insure that their contributions have disproportionate impact? First, we must apply effort where it will do the most good. Second, we must employ appropriate means of applying that effort.

The initiatives the AAG has undertaken are designed to find pressure points and to develop levers that will advance the discipline. In fund raising, geographical education, and publicity, the AAG is bending every effort to get geography's message to people who decide matters that are important to geography, and to make sure the message is presented by the discipline's most effective spokespersons.

Every geographer or group interested in advancing the discipline must pursue similar initiatives locally.

1. Administration is one way to multiply the discipline's influence. Good work done by geographers creates favorable impressions even if the work itself is not geographical. Any group of geographers that does not take pains to make sure that some of its members participate in administration (and in campus governance if it is an academic group) is overlooking a point of leverage.
2. Specialization is another way to exercise leverage. Research and instructional centers (focusing on African geography, remote sensing, or retail location, to pose three possible examples) could concentrate effort that may now be dissipated by dispersion. Educational and research programs have become more specialized in recent years, and I think their quality has improved accordingly. Should that trend result in nationally recognized centers, their impacts would be disproportionate to the number of people involved.
3. Publicity is a third method of getting more out of the work geographers do. Presenting research at meetings, publishing findings in scholarly, applied and popular journals, and speaking to civic groups are all means of informing the larger community of the value of the work geographers do. Geographers are prone to undervalue their work, and the discipline could greatly enlarge its constituency by taking deliberate measures to insure that its successes are trumpeted.
4. Accomplishment is the true lever of success. Taking care that members of the profession are well placed administratively, promoting productive specialization, and promoting accomplishments cannot substitute for quality work. But geography's small size and diversity require that the first-

rate education, practice, and scholarship that we must take for granted be augmented by extraordinary measures to prize the maximum benefit from that work until there are many more geographers than there are now.

Even a single rabbit could move the earth with a proper fulcrum and a long enough lever. With deliberate attention to finding fulcra and creating appropriate leverage, geographers, despite their small numbers, could mold the earth nearer to their hearts' desires.

I. M. MOODREE (GEORGE DEMKO)
October 1986

THE REGULAR COLUMN BY THE AAG PRESIDENT is being replaced this year by letters from a friend of G. J. Demko. These letters provide insights about the profession from the perspective of a non-geographer. Letter #2 follows:

Dear George:

I applaud your determination to address the Association's finances. Every plan or idea to strengthen an organization or profession depends on resources—geld, dough, moola. Nothing's free or if it is, it's dangerous or addictive. Your plan to develop an endowment fund surprises me only in that it had not been implemented years ago. Let's hope that geographers are generous enough to contribute to something which is in their own best interest and that their loyalty is strong enough to have them include the AAG in their Last Will and Testament.

Speaking of loyalty, before I left the country I had the opportunity to loiter around a few geography departments. I was greatly chagrined to find that a number of your brethren have a Peter the Apostle complex and deny an affiliation with the venerable and ancient discipline of geography. I heard trained geographers claim to be economists, demographers, and even urban ecologists! I had no time to count denials before the cock crowed, but it's clear that some of your merry band are defectors (traitors?). I'd hate to think what they might do for thirty pieces of silver.

Touché! The Meinig book (The Shaping of America: A Geographical Perspective on 500 Years of History, Vol. 1, Yale University Press, 1986) is superb. The reviews from the New York Times, *the*

Washington Post, *and the* Los Angeles Times *were interesting, and I deserved the tweak from you. It does bring home the point, however, that geographers need not multiply like rabbits to assure the profession a place in the sun. Quality, not quantity, is the key to any profession's survival. It would be more rational and less salacious to get the present group more active and committed. Two more Meinigs and a place to stand and there's no telling what you might move. What can be done now, however, is for every one of your members to buy a copy of* the Shaping of America *(which will make Yale University Press more pro-geography) and review it for their local paper.*

I look forward to more information on your "Task Force for the Advancement of Geography" and your plans for plenary sessions at the Association's meetings in Portland. Fight the good fight, keep well, and remember that old Ruthenian proverb—"If everybody likes you, you must not be doing anything!"

I. M. Moodree (George Demko)
March 1987

THE REGULAR COLUMN BY THE AAG PRESIDENT is being replaced this year by letters from a friend of G. J. Demko. These letters provide insights about the profession from the perspective of a non-geographer. Letter #7 follows:

Dear George:

I have traveled south, over the roof of the world, to the birthplace of the Buddha—the Sage of the Sakhyas. He taught that all of life was pain—that is certainly something you might readily agree with this year! Bear up old friend, nirvana awaits you.

You do know that I am really getting into this travel stuff. Consequently, I am always looking for good information, reliable guides, useful maps, and related materials. Again, I wonder why your organization doesn't do more in this realm. It seems to me that there are a plethora of useful and financially rewarding projects your precious AAG could sponsor or develop. Why not an AAG World Excursion Service? A set of training seminars could be developed for travel agents on a continuing basis and also to travel associations. A regular consulting service would be relatively easy to develop along with booklets, guides, maps, etc.

A natural project for your Association would be the creation of a set of regional, geographic guidebooks written as substantive and accurate documents but in a readable and breezy style. These could be done for the US, Europe, Asia, and the other parts of the world, and marketed through a major publisher. What more authentic imprimatur could such a set of guidebooks have than the Association of American Geographers? If you did it right, the proceeds could fund a chair of geography or two!

In this same vein, and with a little entrepreneurial skill, your Association might develop a radio or even a TV series called "Where in the World?" Or, "Geography for Those Who Mistakenly Think It's Place Names!" Or, "Sophisticated Geographical Analysis for University Deans and Presidents!"

A natural combination would be a cooperative project with the National Geographic Society where they already do super things with the media. A jointly developed set of programs could remove your anonymity and financial distress. What about, "Human and Physical Landscapes of North America—A Series?" or, "Exploring the World Via the Great Maps of History!" Or, "Politics, Plagues and People—A Geographical Perspective." The latter could be a spatial tour de force of maps and analyses of terrorism, communism, AIDS, and refugees.

I grow weary in my efforts to prod you and your tribe to glory. I proceed eastward again and will stay in touch. Until then, may the wisdom and serenity which originated in this place fill your mind.

TERRY G. JORDAN: *Critical Mass*
March 1988

WHY IS THE DISCIPLINE OF GEOGRAPHY so poorly regarded in certain elite academic circles that troubled departments are abolished rather than improved? What explains our banishment from the Ivy League schools, Stanford, and some other prestigious universities? Why does the intellectual establishment in this country pay us relatively little heed? I do not pretend to have the complete answer to these critical questions, for the issues are complex, but I suggest that part of the problem is based in a low level of scholarly book production. Geographers apparently do not write nearly as many scholarly books as professors in the other social sciences and humanities, both in absolute numbers and on a per capita basis. Books, particularly those published by university presses and certain commercial houses offering high-quality lines, are the most overt, accessible expression of the intellectual achievement and merit of an academic discipline, and in this endeavor we seem to fall short.

Admittedly, the device I used to compare the critical mass of scholarly books in geography with those in neighbor disciplines may be flawed, but I suspect the margin of error cannot explain the massive discrepancy it revealed. Taking the period 1976–1985, I counted the book reviews in our two journals. The *Annals* and *The Professional Geographer,* then compared the total to that derived from like journals of neighbor disciplines. I will use anthropology as a comparative example, since that field shared our integrative character, physical/cultural split, and small membership base.

In the decade mentioned, some 2,828 book reviews appeared in *The American Anthropologist,* roughly six times the 460 published in our *Annals.* Even if we added the 1,215 reviews published in the *PG* during the same period, our book total still falls far short of the anthropologists' effort, and we must bear in mind that many textbooks and other non-scholarly publications are reviewed in the *PG.*

Does a "book gap" exist? Is our critical mass of scholarly books deficient? Can our poor intellectual image be related to such under-productivity? I fear the answer is affirmative. Higher education, particularly at the level of the major universities, is a Darwinian ground, and we must ask ourselves whether we are as intellectually fit as we should be. Surely all academic members of the Association have at least one scholarly book in them. Geography very much needs such works, and soon.

Terry G. Jordan: *Anti-Intellectualism*
June 1988

I CARRIED ON AN EXTENSIVE CORRESPONDENCE in my year as AAG president, and some other geographers conveyed their feelings to me indirectly. The unhappiest discovery I made through such communication is that anti-intellectualism is not all uncommon in our field, even among academic members.

Read over my shoulder, if you will, particularly concerning our two scholarly journals. "Many of us would object to giving another cent to them." "Publishing a better journal will not promote our cause." "I express my strong oppositions to any diversion of funds or resources into publications." "Precious little of the mass of research produced by geographers in the *Annals* or *PG* in the past ten years is…very important." "Elementary and secondary schools should be the first priority of the AAG." "Few even read the *Annals* or the *PG*." "[Geography Awareness Week] probably did the discipline more good than perhaps a hundred or so journal articles" (comment published earlier in the *Newsletter*). "Most members now consider *Annals* to be an almost totally useless document" (from the chair of a doctoral-granting department, no less!). "We already possess a body of theory, concepts and findings" and therefore an "increase in funding for research and journals is less important . . . than building geographic awareness and promoting the discipline in the arenas of education, government, and public policy . . . " (from a resolution adopted by the Middle States Division and read into the minutes of the AAG Council at Windsor).

While such anti-intellectualism may actually serve to advance the applied and pedagogical profession of geography, it will, if allowed to dominate the AAG, almost surely mean the death of the academic discipline of geography. Our disciplinary soul will not be saved by Madison Avenue promotions, not by twenty million turned-on adolescents, nor by legions of diligent applied geographers and technicians in the workplace. Rather, we will be permitted to retain our seat among the pure arts, sciences, and humanities only by continuously and more abundantly demonstrating our worth as scholars and intellectuals. A year's correspondence has convinced me that many of us do not even desire that seat for geography. In my mind rises a dark thought: can the murderous deans have been right after all?

Saul B. Cohen: *The Nineties: Opportunity to Attract a New Generation of Intellectual Elite*
February 1990

Predicting what the nineties will bring is the current national pastime. We can probably look forward to a decade of increasing peace and laying the foundation for a genuine world family of nations. Nationally the nineties should mark the end of America's growth through borrowing. Painful economic adjustments can be cushioned by maximizing the existing human and material resources base, and by reducing the excesses in blatant consumerism. The peace dividend that will be generated by the Cold War's end will certainly have an important monetary impact. However, an even greater impact of this dividend may be the redirection of national political energies away from superpower competition, towards the complex problems of our huge and growing underclass, as well as the repair of our degraded environment.

The nineties also hold out the prospects for marked changes in individual goals, values, and behaviors. If the eighties were the "era of greed" and short-term gain, the nineties could be the "era of sharing" and of deferred gratification. The focus, indeed almost the obsession of the university student of the past decade, to enter such financially lucrative fields as finance, banking, law, real estate, and high-tech sales and management, should begin to shift back to teaching, science, the medical and other helping professions, government, and social services.

Such a shift in individual values could have especially positive impact upon the academy, with university teaching careers becoming attractive to much larger numbers of the educated elite. Indeed, we might look forward to the kind of flowering of academically-based intellectuals that took place in the era that immediately followed the Second World War.

If the nineties bring an academic teaching renaissance, we in geography should begin to plan our role within it. As with most university fields, substantial numbers of position openings can be expected by the mid-1990s. So this is the time to reach out actively—to seek a new generation of young men and women, racially, ethnically, economically, and socially diverse, which will provide the intellectual leadership of the twenty-first century. My reference is not so much to larger numbers, as to quality. We will have a chance to compete for the "best and the brightest" as the changing times redirect their attentions from today's popular professions back to the academy as a satisfying and socially useful career focus.

How can geography compete successfully with better-known university-based disciplines? By conveying the potential of the field to link exploration of place with exploration of thought. Spatial frontiers are never filled, they are simply transformed. The art and the science of this transformation can have great appeal to the inquiring mind and personality.

If one hundred geography departments took upon themselves the challenge of recruiting for graduate studies 200 of the very best students in their institutions each year for the next five years, think of the potential impact of such a presence upon our graduate programs. But recruiting doesn't mean public relations or glib persuasion, it means a careful, thoughtful, and time-consuming effort. A national undergraduate honors fellowship society might serve as an organizing vehicle. Some college departments already have such programs, and certain AAG divisions and state organizations recognize excellence through awards—so we have a start.

A comprehensive structure at a college should include a junior/senior honors seminar and the requirement that participants have the equivalent of dual majors. Participants would benefit from undergraduate research fellowships and from being linked to advanced graduate student mentors. Whatever the nature of the college-based structure, the focus should be upon intellectually broadening experiences. Content and techniques should remain within the province of the on-going course structure.

Each year, approximately 130 doctoral degrees are awarded in our field. If half the recipients could ultimately emerge from some sort of process that attracts and demands students of exceptional intellectual talents, geography will be able to face the twenty-first century with confidence and optimism.

THOMAS J. WILBANKS
November 1992

IN AN EARLIER COLUMN, I suggested that the rising external demand for geography offers us intellectual challenges as well as managerial ones. Let's return to this issue, because it lies at the heart of what the AAG stands for.

For the sake of argument, at least, consider the possibility that society in the 1990s is asking some things from geography in terms of content as well as style that are different from what many of us have been doing. If so, what do they want from us?

Well, "they" are not monolithic, of course, and each of us interprets the messages through his or her own filters; but here are some of the things I think I'm hearing, based on such sources as the NAEP public hearings and consensus process:

1. Our society would like us to connect our research agendas more explicitly with important social issues. This does not necessarily mean a shift away from basic toward applied research, narrowly defined. For two decades or more, other fields of research (such as the materials sciences) have recognized the importance of "issue-orientated fundamental research," and our own discipline has been shaped by this kind of concern since the late 1960s. In fact, most of the geographers elected to the National Academy of Sciences since 1970—when we had no members at all—have had in common their contributions to issues

that transcend not only the paradigms of geography but also the concerns of academia.

2. As I see it, however, we have had four problems as a discipline in this regard. One is that many of us have tended to bend only slightly toward the issues of the day, and in the process we have missed some of the big questions. In the 1960s, 1970s, and 1980s, for example, we had astonishingly little to say about avoiding nuclear war, the preeminent social issue of that generation. A second problem is that those of us who have tried to be issue-oriented have often tended to follow the issues rather than to anticipate them. When we wait until the policymaker asks the question to start our research, we have very little hope of basing our insights on strong intellectual foundations. A third problem has been that, even when we have had good work to offer, we have too often tended to talk to each other rather than to people who might use what we might have to say to make a difference. A fourth problem in some cases has been that we have sometimes tried to assert our right to be heard by referring to our disciplinary label, geography, rather than depending on the strength of our insights to carry the day with audiences who care very little about disciplinary labels. Whatever the reasons, the apparent gap between what society (and geography education) want, and what we do is both a challenge and an opportunity.

3. Our society wants more insights from us about nature–society issues. Whether the question has to do with global environmental change or local waste management, our fellow citizens—even down to elementary school age—now understand that our quality of life depends on a balanced relationship with our natural environment, and they are looking to geography for insights about such things as environmental opportunities and limits, environmental perspectives, and implications of technology. Partly, I suspect, this is because the National Geographic Society has shown a strong interest in these issues without becoming politicized (not an easy thing to do), but partly it is because we ourselves have said for generations that this is one of the things we do well: bringing the physical and human worlds together in our studies. The truth, of course, is that we have said it more often than we have done it. Because of the ghosts of determinism, we have often shied away from talking about effects of the physical environment on human action. For reasons that I have never been able to understand fully, physical and human geographers have seldom worked together

very effectively on nature-society research projects; from my own perspective, we especially need more contributions to these projects from physical geographers. And so on. But the main point is not to blame ourselves for what we have not done but to suggest more attention to what we can do, maybe in the process bringing added coherence to our discipline.

4. Our society wants us to show leadership in viewing diverse phenomena in an integrated way. It seems to me that people in the United States are tired of scientific reductionism, even if they don't use those particular words. They know from their personal experience that dissimilar phenomena can be related in important ways: e.g., trees in the Amazon and climatic anomalies where they live; spray aerosols, the ozone layer, and risks of skin cancer; political reform in Eastern Europe and money for social services in America. They ask: Why doesn't somebody pay attention to integration and synthesis as well as analysis? Whose job is it to pay attention to wholes as well as parts? Meanwhile, as a science (at least), we have moved rather forcefully away from the idea that geography aspires to integration, which seemed to lead toward encyclopedic description and taxonomies—or else to impossible intellectual challenges. Maybe it is time to revisit whether we can break new ground in studying relationships between parts and wholes, because somebody needs to do it and no one's intellectual traditions are as close to this as ours.

5. We are being asked not only to contribute to new capabilities for collecting and using geographic information but also to clarify what our own particular disciplinary contribution will be. All of us know about GIS these days, even if it still seems remote to the computer-illiterate; and we all know that the explosion of interest in this explicitly "geographic" tool is the result of rapid (and apparently still accelerating) technological changes in our ability to access, store, display, and analyze geographic information. We know that the GIS movement has spread far beyond the boundaries of geography, if it was ever ours to start with. But we also know that no one is quite our match when we focus our attention on this territory, whether on techniques or applications.

Our problem is that we are presented with an opportunity that is too big for us: We simply have too few geographers available to claim all the territory that might otherwise be ours. As others expand to fill the territory we leave unclaimed, and we are losing ground as you read this column, it is more and more important for us to focus our efforts on what we do best. Personally, I suspect that our strengths as a discipline lie in the content parts of GIS systems—connecting them with important questions and uses, not just with options for generic processing and display. But whatever direction we pick, we should be determined to be the leaders in some important part of the enterprise, not just as individuals but as a discipline.

Obviously, this list of themes is only suggestive, as a basis for discussion. But the importance of these kinds of questions has led to a proposal by the US National Committee for the International Geographical Union for a new appraisal of geography by the National Academy of Sciences/National Research Council as we look forward to the new century. The last, *The BASS Report on Geography*, was issued in 1970, a few years after the publication of *The Science of Geography*, and many things have changed since then, both inside geography and in the world around us. Where should we be headed now? What resources do we need, especially as our research becomes more dependent on such things as technology use and travel support? As we grow, how will we relate to other disciplines and to multidisciplinary research themes? How will our scholarship relate to our teaching, from kindergarten to college? Stay tuned. It's an exciting time for geography—maybe in this intellectual connection more than any other.

LARRY BROWN: *Continuity and Change*
June 1997

REDISCOVERING GEOGRAPHY: *New Relevance for Science and Society* (National Academy Press, 1997) is "a comprehensive assessment of geography in the United States" (p.vii), carried out by the National Research Council, the first such report since the 1970 Behavioral and Social Sciences Panel's *Geography* (Taaffe, et al., Prentice Hall). *Rediscovering Geography* is not an isolated event, but the culmination of public realization in the 1980s that our citizenry suffered from geographic illiteracy; establishment of state-based Geography Alliances by the National Geographic Society, beginning in the mid-1980s; national legislative action instituting an annual Geography Awareness Week and later, *Goals 2000: The Educate America Act*, which included geography as one of the nine core subjects; publication of *Geography for Life: National Geography Standards 1994*, a guide for K–12 curricula; and now, *Rediscovering Geography*. Its foreword by the NRC Chairman states, "The discipline of geography has been undergoing a *renaissance* . . . " (p. vii, emphasis added). In a more boosterish tone, Neal Lineback's recent *Geography in the News* (#400, 4/11/97) is titled "Hot Geography" (see p. 8) but the column quickly settles down to provide a must read account of contemporary geography that should be shared with many. My own columns and President's Plenary Session have been a part of this movement, with the repeated theme of Continuity and Change. So, taken together, but especially focused by *Rediscovering Geography*, a *"bridge to the twenty-first century"* has been established. The rediscovery theme represents a reverberation from several forums. Public recognition of geographic illiteracy through the popular press, business, and government resulted in actions to offset that. Among these, rediscovery took especially tangible form in the K–12 curriculum and the College Board's Advanced Placement program. Student enrollments at all levels have increased dramatically, both in course-takers and majors, an increase that well outstrips the experience of other disciplines over the past decade. In research, other specialties have increasingly discovered, and now even emphasize, the geographic perspective, the importance of space—"extending the influence of geography well beyond its relatively small group of professional practitioners" (p. 9). Geography also has been rediscovered in the market place. Even though *Rediscovering Geography* takes a cautious, conservative view, it nevertheless states we have a distinct comparative advantage in offering an education that "combine[s] technical skills with a more traditional liberal arts perspective" (p. 217). Geography curricula provide numerous transferable skills, an increasingly important spatial perspective, and ability to effectively utilize the burgeoning array of spatially referenced data. Because of this, and recent AAG efforts focused on the non-academic market place, employment opportunities for geographers will expand considerably. Finally, rediscovery occurs by society's recognition that solving important problems requires a multi- or interdisciplinary approach, and geography is one of the few disciplines that incorporates this perspective within its bounds.

    Rediscovering Geography illustrates the discipline's relevance by application of its tenets to critical issues (Ch. 2) such as economic health, environmental degradation, ethnic conflict, health care, global climate change, and education. Chapter 3 elucidates the geographic perspective, emphasizing the role of place and scale, physical and social science, and representation. Chapter 4 depicts the range of geography techniques—field observation/exploration, remote sensing, cartography, GIS, visualization, and spatial statistics. An important message is that these techniques represent transferable skills which are relevant to endeavors other than the purely geographic. Chapters 5 and 6 elucidate geography's contributions to scientific understanding and decision making. The role of place, scale (local, national, international, global), and geographic representation is emphasized; examples of decision making include urban policy, water resource management, retail marketing, legal dispute resolution, energy policy, technological hazards, flood plain policy, information infrastructure, global environmental change, economic and political restructuring, technology–service–information transfer, and hunger. Chapter 7 addresses the need to strengthen foundations in issues such as complex systems, global change, the global–local (macro–micro) continuum and shifts across scales, cross-national comparative analyses, and geography education.

Each chapter has numerous examples (sidebars) which, taken together, indicate the breadth and diversity of geographic inquiry. These include 'Spatial Diffusion and Epidemics" (pp. 49-50); "Climate and Vegetation Change" (p. 59), "Interactive Analysis of Spatial Data" (pp. 66-67); "Civil Unrest in Los Angeles" (pp. 78-79); "Urban Climatology" (p. 80); "Land Use and Soil Erosion" (pp. 84-85); "Long-Wave Rhythms in Transnational Urban Migration" (p. 94); "Food and Famine in the Sahel" (p. 100); "Future Geographies" (pp. 106-107); "Urban Policy: Housing in Minneapolis-St. Paul" (p. 113); "Management of the Colorado River" (pp. 115); "Dispute Resolution: Electoral Redistricting in Los Angeles County" (pp. 118-19); "Adaptation to Climatic Change in Major River Basins" (p. 128); "Interactive Tools for Geographic Instruction" (pp. 156-57).

The book closes (Ch.8) with recommendations related to "Improving Geographic Understanding and Literacy," "Strengthening Geographic Institutions," and Taking Individual and Collective Responsibility for Strengthening the Discipline." The last issue is especially pertinent:

> *"Geography's new relevance does not just pose challenges to external bodies and larger institutional settings; it calls for response by individuals and groups of geographers . . . [who] need to recognize they also have responsibilities to their discipline, to other sciences, and to society" (p. 166).*

Our actions must not, in my opinion, reflect individual interests more than (rather than?) disciplinary citizenship, especially because it takes only one disaffected, self-interested geographer in a critical place at a critical time to erode our position in the academy and society. In this regard, as AAG President I have been buoyed by the number of geographers who are extraordinarily, incredibly loyal to their discipline, enraptured and energized by the things we do—but vigilance and non-complacency must be maintained. Let us not have to repeat after Pogo, "We have met the enemy and he is us."

Every geography unit should have a copy of *Rediscovering Geography* to share with (give to, if feasible) relevant administrators, faculty, and persons outside the academy. Likewise for *Geography for Life: National Geography Standards 1994* and Neal Lineback's "Hot Geography" column referenced above. Several of you have found it useful to share my president's columns; I encourage others to consider this, and will gladly provide a set on page form, separate from the *AAG Newsletter* itself. Columns that apply directly to *Rediscovering Geography* include July on, "What Do Geographers Do Anyway;" September on "The 'G' in GIS;" December on "Geographers in Business, Government, and Non-Profit Organizations;" April on "GeoEd."

It has been a real privilege, and lots of fun, to serve the AAG as president. To end my term with the opportunity to highlight *Rediscovering Geography* and other bridges to the twenty-first century is an undeserved capstone. Geography is healthy, well, and prosperous. Thank you all for the opportunity given me.

REGINALD G. GOLLEDGE: *NEVER Be Ashamed of Being a Geographer* June 2000

THIS IS MY LAST PRESIDENTIAL *NEWSLETTER* COLUMN. I thought often about potential topics. Finally, after the Pittsburgh AAG Council Meeting had been completed, I was exchanging ideas with Bill Dando (West Lakes Council Representative) when I had occasion to speak the words that I selected as the title for my last presidential column. I agree heartily with this sentiment.

Over the years, I have seen a great deal of denial of disciplinary roots by geographers. Some just refuse to call themselves "geographers" and use titles like "earth scientist," "statistician," or "environmental engineer." Some refuse to acknowledge their departmental home on their publications, instead citing research units or other institutional bases. Some feel that geography is so poorly regarded by other disciplines that they cannot compete equally for fellowships, scholarships, and research grants and contracts if they identify as geographers.

I've always wondered about this reluctance to admit to a geographic heritage, more so lately as more geographers obtain million-dollar grants and contracts in open competitions with other science, engineering, and humanities consortia. I think I've always been a geographer, and though my ideas about geography have changed substantially over the decades, I'm still proud to be a geographer. I respect the discipline for providing a knowledge base that can be extended into many problem domains. I respect the discipline for providing me with an academic home that has allowed me to pursue research goals that—at times—must have seemed very peculiar to other geographers. I respect the profession that has so much to offer to environment and society, individual and policy. And I cannot for the life of me understand those who reject their geographic heritage and state that they are ashamed of being a geographer.

Our discipline has always faced the challenge of competing with other larger, more powerful, and more widely accepted disciplines—particularly in the US, although less so in many other countries. Often, representatives of other disciplines disparage geography. But this attitude is frequently based on ignorance—ignorance of the essential part geography has played in the development of civilization over time; ignorance of how large a part geography plays in our everyday activities; ignorance of the nature and topical areas of today's geographic researchers; ignorance of the additional strength that a geographic perspective can add to almost any research area that concerns the interacton between humans and environments.

There have been times, of course, when groups have found themselves disenchanted and at odds with prevailing concerns—usually because they are outside the mainstream. Such groups usually do not accept that what is represented as geography at that time is the same as the geography that they support or practice. Historically, these times have been turbulent, replete with angry criticisms and disclaimers. But often they enrich the discipline by introducing new theory, new data, new perspectives; by spawning new journals; or by overcoming barriers to new interdisciplinary contacts.

I believe that geographers who admit to being ashamed usually are those who do not understand the discipline, or their place in it, and consequently have been unable to convince themselves that geography has unique strengths to offer to the world of knowledge acquisition. I also think that, if a few solid home truths about our discipline became more widely accepted, those feelings of shame would erode. Being proud of our discipline and being prepared to enlighten others about our unique contributions should help build up disciplinary confidence—the lack of which appears to be the foundation of shameful feelings.

Recently, Goodchild (2000 AAG Meetings) identified a number of key concepts that help differentiate geographic research from other disciplinary emphases. I take some liberty in rephrasing and supplementing this list as follows:

1. *Geography is an Integrating Science.* Process theory and models and methods that span and link physical and human environments and activities provide a common base for geographic reasoning. This type of reasoning is peculiar to geography.
2. *Spatial Analysis.* Geography has developed a unique way of examining spatially referenced data that is not duplicated by any other discipline. After decades of slow but sure diffusion, spatial analysis is an investigative mode that is becoming sought after by many of our sister disciplines.
3. *Spatial Representation.* Geography had perpetually been associated with mapping. The use of maps, graphics, and other image representations facilitates what psychologist David Uttal describes as a unique way of learning about places and human-environment relations that is not duplicated by any other learning modality.
4. *Spatially Explicit Theory and Models. Geography* has produced a set of spatially explicit theories and models that represent unique perspectives for understanding the distribution of phenomena and the processes that produce those distributions. In particular, these theories and models stress that changing the location phenomena changes the ability to forecast, predict, or explain their relations.
5. *Place-Based Analysis.* Much of geography's power lies in occupying the middle ground between nomothetic (generalized) and ideothetic (specialized) reasoning that allows place-to-place variations to be recognized, analyzed, and represented in forms ranging from comparative local area descriptions to the use of models whose parameters are place-specific.
6. *A Unique Way of Combining Knowledge and Policy.* Geography facilitates the combination of knowledge of local and global conditions and societal needs to help develop and implement place-specific policy.

7. *Place-Based Search.* This allows the compilation and use of spatial and non-spatial information relevant to understanding a geographic footprint, such as by collecting the map, diagrammatic, photo, video, satellite imagery, or other records that might help to help solve place-based problems—as is being done via the Alexandria Digital Map and Image Library.

8. *Scale.* More than any other discipline, geography concerns itself with the effects of variations in scale in both physical and human domains—effects that often are assumed away or ignored by other disciplines.

Each of us can elaborate on these concepts with anecdotal references to our own research and teaching, or to the work of our colleagues. There is no shortage of evidence to support the above claims that geographers offer a different (and needed) perspective. When traveling to the different AAG regional meetings, I've observed posters and heard papers from students, faculty, and non-academics that attest to this. And, shortly, we will try to implement a Geography Research and Education Network that will make many examples of the variety and quality of geographic activities freely available at the touch of a keyboard or the click of a mouse button.

To be rid of any image of shame, we must be prepared to convince other disciplines that we have much to offer. Recent projects like NCGIA and Project Varenius, Global Changes in Local Places, ARGUS and ARGWORLD, and ongoing projects like Mission Geography and Thinking Spatially are showing academic, teaching, and general public audiences that geographers need not be ashamed of the things they do. But we must continue to enlighten other disciplines by showing—through quality academic papers published in our journals, by successfully competing against them for National Centers and Institutes, by undertaking large and serious projects that have visible impacts on knowledge acquisition—that our discipline has unique capabilities and insights that both complement and transcend the work of others and, thus, cannot be ignored.

As concepts of environmental sustainability and vulnerability enter more frequently into even casual conversation, and as "integrated science" is more emphatically seen to be the most appropriate way to do science, we have a unique opportunity over the next decade to build on successful endeavors of the recent past and to mould geography into a discipline that even the naysayers cannot disparage. Let's do it!

SUSAN CUTTER; *Bring Geography back to Harvard and Yale and...*
October 2000

THE LACK OF FORMAL GEOGRAPHY (courses, an undergraduate minor, major, or graduate study) in many of the most prestigious universities in the nation is a missed opportunity for these elite institutions of higher education. There are leaders in American society—business, politics, law—with no knowledge of the discipline's contributions to understanding and solving contemporary social, economic, or environmental problems. It's not that they "dislike" geography or have had bad experiences with it (although both could be true). Rather, many of the nation's contemporary leaders have never been exposed to the discipline at all. The elite educational track—private prep schools, Ivy league colleges for baccalaureates, professional degree programs (law or business)—simply does not have geography in the curriculum. As a consequence, there is a gaping educational and intellectual void in graduates from elite universities. How can the elite schools claim to provide a scholarly education when one of the most fundamental and core disciplines, geography, is absent from their curricula?

Historically, geography was in the curriculum at many of the nation's leading institutions—Harvard, Yale, Penn, Princeton, Columbia, Brown, Duke, and Stanford. Before 1800, geography had a prominent place in the curricula of colonial colleges. It was institutionally placed among the other science subjects—astronomy, chemistry, mathematics—and emphasized physical geography and Cartesian coordinate systems. After independence, the rise of nationalism and the need to teach children about the newly formed nation, the orientation of the discipline changed from a systematic natural science at the global scale to a descriptive regional focus at the national level centered on K–12 education. Colleges were established to train teachers, and geography became identified as the subject matter for K–12 teachers in the normal schools, not a subject for serious students in major colleges and universities.

Geography began to re-establish itself following the Civil War, when Arnold Henry Guyot held the Chair of Physical Geography at Princeton (1854-1884) and Daniel Coit Gilman became a professor of geography at Yale (1863-1872). Harvard became the focal point for training physical geographers (in a geology department under the direction of William Morris Davis, AAG founder and first president. The first independent department of geography offering geography degrees through the PhD was founded in 1903 at the University of Chicago with Rollin D. Salisbury as chair. At Harvard, geography assumed the status of a department within the division of geological sciences in 1907. With Davis's retirement from Harvard, Wallace Atwood became chair. In 1920, Harvard awarded its first PhD in geography, but Yale beat them to it—awarding its first two PhDs in geography in 1909, to Isaiah Bowman and Ellsworth Huntington. Yale also has the distinction of awarding the first geography PhD to a woman (Gladys Wrigley in 1917). Atwood left Harvard in 1920 to become president of Clark University where he created the Graduate School of Geography (still a thriving program).

Over the ensuing years geographers have taught and awarded degrees in many of these elite institutions, but there now exists no sustained geographical presence, with the exception of Dartmouth and the University of Chicago. Dartmouth continues as a strong geography program today, the only autonomous department of geography among the Ivy League schools. The University of Chicago maintained a strong and top-ranked department of geography until the mid-1980s, when it was downsized to a committee status.

Geographical knowledge is more important than ever in an increasingly global and interconnected world. How can a graduate claim to be a learned scholar without any understanding of geography? As I mentioned in my last column, geography has achieved notable recognition for its role as an integrative discipline (in global change research and in information science) and for its regional perspectives in understanding global interdependence. The field is advancing rapidly with theoretical, methodological, and conceptual innovations regarding the delineation, demarcation, and representation of space and the underlying processes that give rise to interaction, connectivity, and distribution of phenomena within it. We approach our subject from many different systematic subfields and with many different lenses, yet we are all fundamentally concerned with place, space, and location. Given the rediscovery of geography by many institutions ranging from the National Academy of Sciences to the addition of human geography to the College Board's Advanced Placement (AP) program, isn't it time that the elite universities and colleges also rediscover geography and stop lagging behind state universities and colleges? Of the myriad of reasons why geography needs to be reinstated into the curricula, I will focus on only a few.

First, with the rapid advances in information technology, geography is uniquely suited to develop the requisite theoretical underpinnings, foundations, and subsequent applications of geographic information science to a host of contemporary problems and fields such as business, natural resources management, public health, and urban planning. Harvard launched academic GIS through the basic research of Howard Fisher at the Harvard Lab for Computer Graphics and Spatial Analysis, yet the core discipline that teaches about computerized applications of geographic knowledge is not present on that campus.

Second, geography is a liberal arts discipline and should be part of a liberal arts education. It spans the traditional distinctions of natural science, social science, and humanities, since all these perspectives fit under our umbrella. As a field of inquiry, geography provides a broad-based understanding of where phenomena are located, why they are there, and how they relate to other phenomena in that place. Geography fundamentally addresses the nature and significance of patterns, be they the interactions between cities and their suburban communities; the unequal distribution of environmental threats; the replication of physical processes that create different landforms; the evolution of cultural landscapes or the diffusion of internet viruses. Geography provides a framework for interpreting facts and a way of organizing them to make sense of the world.

Third, geography provides an understanding of places and regions and the interconnectedness between them ranging from the local to global scales. Understanding the basic pattern and processes of landscape change positions geographers to contribute productively to formulating policy in realms as varied as biodiversity protection, social welfare reform, industrial restructuring, and global trade.

There is no overall conspiracy to keep geography out of the elite colleges and universities. Faculty and administrators at many of these institutions accept geography and are willing to hire individual geographers. What geographers need to do, however, is convince one or all of these prestigious institutions to reinstate geography as a degree-granting program. Let's bring geography back to Harvard (or Yale, or...) where it belongs.

Note: I am indebted to Allen Bushong for his help in understanding the historical background regarding many of these departments. I look forward to receiving your thoughts on how we might convince these elite instructions to add geography to their curriculum and geographers to their faculty.

❋ ❋ ❋

Chapter 12

ETHICAL ISSUES

❋ ❋ ❋

For many AAG Presidents, the question of ethics—both in our science and in our practice—was a topic for one or more columns. These ethical issues are both internal and external to the discipline and range from our roles and moral responsibilities as professionals to ethical issues surrounding the use of geographical technologies.

❋ ❋ ❋

ROBERT KATES: *What Will You Do When Hate Hits Your Campus?*
October 1993

OUR HOUSEHOLD RECEIVES the *Intelligence Report* of the Southern Poverty Law Center. Every two months, the Report lists a sampling of hate crimes in the United States. In the most recent issue, I counted six murders, three arson attacks, forty-seven assaults, eleven cross burnings, two weapons violations, six threats, thirty-nine cases of vandalism, six clashes, and twenty-one cases of harassment and intimidation. In all there are 141 cases—ten of which, from their brief descriptions, occurred on or around college campuses. Hate crimes are common in America and increasingly common on college and university campuses. What do you do when they occur on your campus?

This column is prompted by the agonizing experience of one of our premier graduate departments that found itself ill-prepared to cope with a series of racist and political hate communications and vandalism directed at graduate students and faculty members. Their experience, one that they have shared readily with colleagues who have inquired, has led me to reflect on my own experience. In so doing, I drew upon three sets of encounters; the response to racist, sexist, and homophobic harassment on my own campus (Brown University), the human rights activity of the National Academy of Sciences, and yes, some personal history. From these in answer to the question I posed above, I can think of half a dozen "do's" worth sharing.

Take every incident seriously.

For many of us, our academic settings are comfortable places. Hate intrudes, destroys the harmony of purpose, the expectation of shared values of diversity, tolerance, and mutual respect. We respond with denial, questioning the intent of the incident (indeed there may be some ambiguity), its source (surely from the outside), its importance (if you take it seriously you only encourage the perpetrator). We can most easily get beyond denial by considering the irrelevance of these considerations to the victim and putting the victim first.

Support the victims.

The victims bear the burden of the threat or the assault, the violation of their space and their person. The most important response is to marshal sympathy, support, and solidarity for the victims in big and small ways. Whatever ambiguities of intent, source, or importance might exist, the victim's hurt is real; their need for collegial support should be dominant.

Press for resources.

Colleges and universities have many resources to address hate crimes; to provide security and counsel; to alert the community and dispel rumor; to seek cause and justice; to reaffirm values; and to prevent reoccurrence. But seldom are these resources fully and promptly utilized. Beyond denial, there are institutional inertia, competing campus crises, or fundamental insensitivity which limit the scale and timing of responses. We—immediate colleagues, faculty and students—can serve as important institutional advocates to press for the full and rapid deployment of these resources.

Address the larger context.

Hate on campus takes place within a larger context and one that we should confess we do not fully understand. The increased expression of racial, ethnic, and religious hatred in the post-cold war world should caution us in attributing specific incidents to parochial circumstance or to simplistic explanations of competition, status, material or political gain, or even human nature. But there are clearly institutional settings that encourage the expression of hate or discourage and repress it. At Brown, I was impressed how the clear and unambiguous rejection by its President of campus hate and his vigorous efforts to seek justice and support for the victims created an atmosphere that encouraged the best in community response and discouraged further incidents.

Expect conflict.

The difficult issues of human rights are rights in conflict. Responding rapidly and forcefully to hate expression on campus can conflict with equally held values of freedom of expression or due process. There is a line to be drawn, albeit at times fuzzy and imprecise, between the hate–motivated crime and the free expression of any and all views and ideas, no matter how repugnant. And there is another line between the community's wish to quickly move against hate perpetrators and the protection of individual rights and privacy that good due process requires. A campus community will be frequently divided on where such lines are drawn. My own fuzzy line drawing tries to distinguish between personal harassment that is targeted on individuals and the general expression of hateful views. Harassment I would punish severely, but the right to expression I would defend. Thus I can distinguish difference between spraying racist graffiti on a dormitory room or office door (personal harassment) and distributing generic racist literature in the street (expression). Yet many situations are much less clear. Thus, I sympathize with the struggle others engage in seeking to draw principled lines between these and similar rights in conflict, even if as is often the case, their lines differ from mine.

Seek common ground.

Almost inevitably, the intrusion of hate into a campus community can set its members against each other. Especially in cases where the perpetrators are not identified or there is some ambiguity in the circumstances, the frustration of community members may vent itself internally, directed toward colleagues, leaders, or administrators, rather than toward the unknown perpetrators. Similarly, a community can be polarized rapidly around differential views of rights that are in conflict. Indeed, these may be exacerbated by external pressures and media coverage that either view efforts to prevent hate on campus as political correctness or conversely, the provision of free expression and due process, as racial or gender insensitivity. The tragic irony of course is that in such polarized situations, the hatemongers win by effectively dividing the community.

Thus it should be possible to seek common ground around prompt recognition of the seriousness of hate crime and expression, support for the needs of the victims, mobilization of campus resources, and creation of a climate that discourages harassment. We will find it more difficult to create a consensus of our understanding of the larger context in which we try to balance rights in conflict. In dealing with these honest differences, we need to trust and respect each other's good will and good intentions— a trust and respect that is the essence of collegiality.

Bob Kates
February 1994

As pressures grow on scholars and scientists to perform in both research and teaching, as the temptations to both take shortcuts and to capitalize on professional knowledge increase, and as the lines between advocacy and science blur, efforts to create professional standards to guide such behaviors increase. I sit surrounded by disciplinary and professional standards, that of the political scientists, A *Guide to Professional Ethics*; the anthropologists, *Professional Ethics*; the historians, *Statements and Standards of Professional Conduct*; the sociologists, *Code of Ethics*; and the National Academy of Sciences, *Responsible Science, Volume 1: Ensuring the Integrity of the Research Process.*

We geographers do not possess such a statement, and occasionally, prodded by our Executive Director, we wonder if we should. It's not that we have not tried, for on at least three occasions in the last ten years the AAG has commissioned groups to investigate whether the AAG should develop a code of ethics or practice. I am not surprised by the difficulty in designing such statements. The samples before me cover such topics as the rights of human subjects or research populations (including issues of confidentiality and informed consent); publication and review processes (including plagiarism and bias); teaching and supervision (including issues of responsibility, confidentiality, and respect); employer-employee-sponsor relationships (including issues of discrimination, sexual harassment, and misrepresentation); and public service (including issues of objectivity and partisanship).

Statements vary in their coverage of these topics and in their approach. For example, sociologists and anthropologists both express concerns for their research populations. However, the sociologists argue first and foremost for objectivity and integrity in their research, and then to do no harm to those they study. Anthropologists declare that their "first responsibility is to those whose lives and cultures they study." In a recent court case in British Columbia, testimony by anthropologists was discounted by the judge because of that commitment; the judge held that the anthropologists by their own standards could not be neutral and therefore objective expert witnesses.

Faced with the difficulty of sorting through these topics of concern and conflicting stances, some professional groups have adopted more general statements such as the geologists' recent adoption of the National Academy of Sciences statement on responsible science. But would an AAG statement of ethical standards be worth the collective labor required to create it?

Frankly I do not know. To help me decide, I have engaged in a modest experiment. I looked back on my own career and asked myself about the occasions when I found myself in an ethical dilemma. Here are four from my own experience. They are not unique and should be familiar in some form to many readers.

1. As a graduate student doing dissertation research, I learned that any research interaction with people may cause some harm. Doing surveys of flood hazard perception, I discovered that in rare cases my questions raised anxieties and fears in my interviewees, which, while mostly unwarranted, were certainly uninvited. Indeed, even the actions of the research team in measuring street elevations to calculate flood risk created rumors that led some residents to fear that their homes would be taken for a highway expansion.

2. When a government agent showed up in my office and asked questions as part of an employment check on a student, I found that I objected in principle to the nature of the questions, which probed political beliefs and personal behavior seemingly unrelated to the job qualifications. I wanted to refuse to answer on principle, but worried that my refusal would be taken as evidence of the student's unreliability, that

I was reluctant to answer not because of my beliefs, but because the student had something to hide.

3. A graduate student showed me a copy of a dissertation proposal submitted by a degree candidate at a university in another country. A third of the proposal was taken directly from her own paper without any attribution. She was hurt and angered by the plagiarism and yet when I urged her to be a whistle blower and write to the department in question, she refused.

4. Our department was deeply committed to affirmative action, in particular mindful of the gap between the different proportions of women students and women faculty. With two posts to fill, we engaged in a search, but found only one exceptional candidate, a male. We debated postponing the second appointment to the following year knowing that the pool of women with doctoral degrees and experience was continually increasing. A colleague questioned our proposed action. Would our action, essentially creating a quota, be legal? And even if it was legal, was it ethical? At the very least, wouldn't we have to inform all male applicants, that qualifications being relatively equal, they did not have an equal opportunity for the post?

What characterized each of these dilemmas was that they involved principles in conflict; my own self-interest and societal interest in my research versus the rare but nevertheless harmful anxiety engendered by the research; my own strong, civil-libertarian views and my student's right to have me provide evaluations and referrals; my responsibility to prevent or expose plagiarism versus my student's right to have her wishes respected even if they were misguided; our need to create opportunities for those denied it in the past versus the reduced opportunities for some in the present.

The sample professional statements that I have reviewed do not resolve these dilemmas; if anything they sharpen them by stating the need to honor each of the principles concerned. That by itself might be an important contribution, especially if a professional statement is used as a basis for undertaking what for me is the central task: making these and similar ethical concerns part of the educational process. No graduate student should complete her or his education without being exposed to a lively exploration of these issues; all should receive the mentoring required to guarantee that sensitivity to ethical standards of teaching, research, practice, and public service become lifelong habits.

What should we do now? There exists already a plethora of institutional policies and regulations governing such areas as confidentiality, conflicts of interest, discrimination and harassment, employment, human subject research, scientific misconduct, student rights, and the like. Should we, in addition, try again to develop our own set of professional ethics? Or should we search for some appropriate generic statement and consider adopting it? Or should we develop some teaching materials to be used in graduate "pro", "core", or "research methods" seminars and encourage our students to hone their own fine senses of professional conduct, rights, and responsibilities?

If you have some thoughts on the matter, or even better, some experience in teaching or drafting suitable materials, why not let Ron Abler know in time for the AAG Council to discuss this further at its San Francisco meeting?

STEPHEN BIRDSALL
September 1994

MOST OF THE TIME, I am confident that we are maturing as members of an organized discipline. Now and then I wonder whether we have matured enough.

Occasionally something happens that suggests that too many individual geographers remain professionally and intellectually intolerant and unaware or unconcerned about the indirect consequences of this intolerance. Here is an example:

At a gathering of academic administrators this past year, I praised another administrator's geography program and asked if he was aware of the good things happening in that program. Yes, he was, he observed somewhat thoughtfully. Then he added that it was too bad the faculty were having so much trouble agreeing on anything. The disruptive disputes between the faculty members, he said, seem to be based on differing views of what disciplinary approaches for the program should be emphasized. He also said personalities were involved.

Then he stated: "Any discipline that has trouble defining its boundaries is going to be vulnerable during these difficult budgetary times."

I reminded him that boundaries change, as he knew from his own discipline, that differences between individuals were not necessarily a reflection of disciplinary deficiencies, and that the field's core is much more important in these matters than the boundaries.

I do not share this exchange to raise anyone's blood pressure. As a discipline we too often give such comments more weight than they deserve. But it is representative of the ways individual behavior is interpreted and projected onto geography as a whole, and such projections can be important.

There is an issue here about which we must be absolutely clear. Scholarly disciplines can not remain vital without testing and incorporating new ideas through open debate. But when proponents in these debates stop listening to each other, become too dogmatic, and load the exchanges with personal attacks, administrators who are not geographers begin to question the discipline we represent.

The core traditions of geographic study have been around for many centuries. The questions have changed. Applications have changed. Demands for the insights of this tradition or that tradition have changed. But the basic concerns of our discipline and the traditions that address these concerns from different points on the intellectual compass have persisted for a very long time.

Many geographers pursuing their own professional challenges are more accepting of other approaches than in times past. I like to think that most of us believe there is more than one good way to pursue the issues we find interesting and valuable.

There is a great need for what we have to offer as geographers. Have we matured enough to keep our internal debates open and constructive, and also to know when it is in our best interest to rejoin in common purpose? I sincerely hope so.

STEPHEN BIRDSALL
February 1995

I KNOW A DEPARTMENT that does several different types of things very, very well. Its faculty, however, are deeply polarized, each side having members who actively belittle and undermine the other side, both what the other group does and its members.

This occurrence is not unique in academia. Departments in many disciplines do this to each other, to themselves, and to knowledge itself. Geography, too, has its share of such cases.

Outsiders view this uncivil warfare, when they become aware of it, with puzzlement. One wry observation frequently offered in explanation is that "political battles in academia are so nasty because the stakes are so low."

This is somewhat unfair. Some such battles are very important in the long run; they just don't seem significant to bystanders concerned with other, more immediate matters. And a few disputes are purely intellectual exercises, pushing our knowledge forward through vigorous testing against the resistance of others' views.

But some of these battles are nasty and destructive. Why? What attitudes do people carry into their relations with others such that strife is a frequent outcome?

I think one of the most pernicious attitudes is that which leads individuals to identify their own worth as a function of the relative worth of others. People who act on this attitude enhance their view of themselves, their "value," by working to diminish the apparent value of others. They build themselves up by tearing others down.

Examples of this life-view abound in wars, demagoguery, racial and gender prejudice, economic and political exploitation, and in many, many other forms. It occurs in academia, as well.

I refer to this way of dealing with oneself and others as pernicious because it is so prevalent, so ruinous to human relations, and apparently is present in more subtle ways than many credit it.

I raise it here because it seems to me that this describes too often our professional interactions with each other, geographer to geographer. Fortunately for our discipline, professional relations are generally constructive, or at least more generally so than in the past. But when these relations are not constructive, they undermine our discipline as a whole. It is not just the intended target in a specific exchange who is harmed, but all of geography.

How should we engage in critique of quality, of conclusions, of approaches in a constructive rather than destructive manner? How shall we recognize the latter even when couched as the former?

I think the basic principle is this: Worth is inherent in the potential quality of the project, just as worth is inherent in the potential of the individual. Geographers who produce scholarship valued widely by others (and not just by other geographers), geographers who are recognized by their peers and their institutions as outstanding teachers, geographers who, by means of their service, enhance with great effectiveness the abilities and productivity of our top scholars and outstanding teachers—these individuals have come close to realizing the full potential of their chosen products. All contribute to the common good, and all should be valued.

Scholarship is not inherently superior to teaching, nor is teaching inherently superior to scholarship. There are many flavors of geographic pursuit, each of which can contribute to the broader goal of increasing our knowledge and understanding. Disagreements over the value of this conclusion or that, this approach or that, can be tested directly by producing and offering an alternative. Many conclusions and approaches will not be sustained over time, given active testing, but it is the time that will tell, not the passion with which the position is argued at the moment.

Others' choices need not be less worthy simply because they are not our own. In geography, there is too much to do for us to unwisely diminish the value of others' efforts in non-constructive ways.

Will Graf: *Fakery in the Publications Game*
April 1999

During my service as associate editor of several journals, AAG councilor and officer, reviewer of grant proposals, evaluator of candidates for promotion and departments, I have encountered three errors often made by authors: self-plagiarism, slicing the baloney thin, and deceptive reporting of publications on resumes. Within the last year, I have personally encountered instances of these transgressions, and colleagues report that there are all too many cases of these transgressions in our discipline. Those involved range from junior to senior faculty. I'm not certain whether these errors are intentional or simply made out of failure to recognize acceptable modes of behavior for authors, so in this column I briefly explore these difficult issues in hopes of bringing some clarity to them, or at least stimulating discussion about them among colleagues.

Self-plagiarism

Many journal and book editors have problems with self-plagiarism, the process whereby an author duplicates word for word in a present publication what he or she has written in a previous publication. Given the nature of our research and reporting, it is inevitable that many of us need to review the same issues or techniques in more than one paper, so that a certain amount of repetition of ideas is bound to occur. This repetition can be reduced by citing one's previous publications, and clearly indicating the potential for duplication. Another approach is to fully outline the issue, technique, or literature in the first publication, and then summarize subjects briefly in subsequent related publications.

What is not acceptable is the word for word repetition of many sentences and whole paragraphs from previous publications. This practice is wasteful, unnecessary, and possibly fraudulent. It is wasteful because journal space is at a premium, and space occupied by self-plagiarized material is space that cannot be used by authors wishing to publish original material. Self-plagiarism is unnecessary because by definition, the material that is appearing for a second time is already in the literature and is already available to users of the information.

Equally important, however, self-plagiarism exposes the author to charges of fraud. This issue is in part strictly legal. Most authors do not own the copyright to their published articles. Through formal signature by the author, most journals require the transfer of copyright from the author to the publisher. As a result, if an author duplicates his or her previously published sentences or paragraphs, there is often a problem of copyright infringement suffered by the publisher of the original. Self-plagiarism also makes the author appear deceptive, because if the duplicated material is not cited to its originally published source, the author appears to be trying to "double-dip." Because of the exact duplication involved, the author is usually aware of the transgression. Many geographers in academia and in agency positions have their promotion and merit pay partially linked to scholarly productivity, which is often measured in number of published papers. Self-plagiarized material thus is counted not as a single contribution, but is counted twice, providing a false picture of the true range of contributions made by the author. Self-plagiarism is therefore not acceptable and should not be tolerated by editors, authors, readers, or evaluators in academic institutions, agencies, or the private sector.

Slicing the baloney thin

Many researchers report the results of their work in a few relatively long, major articles that represent substantial contributions to literature and knowledge. Scientists often publish the results of extensive investigations in brief papers or technical notes that are only a few pages long but that condense succinctly the results of a considerable investment of time, talent, and resources. Some researchers publish their results in numerous publications, with each article revealing only a limited portion of the entire story. This process of slicing the baloney thin produces numerous publications that are very closely related to each other and that are often partly repetitious. The result is a literature cluttered with reports small in contribution, if not in page length. Those who engage in this practice have lengthy resumes with numerous citations, but they cheapen their own work by the process, and evaluators of their contributions discover the ruse upon reading the products.

Many academic institutions and agencies contribute to this problem by evaluating their employees simply on the basis of number of publications, and evaluation committees cannot or will not take the time to actually read the published work. Hence the conventional wisdom is that evaluators can count but can't read. Many institutions have recently become more sensitive to this problem, however, and have taken a variety of approaches to exert some control. For example, some evaluation committees will consider the entire list of publications of candidates for promotion or salary increases, but require that candidates designate two or three publications that the committees actually read. In other cases, committees will consider only the five most important publications for promotion or salary adjustment, with the candidates making the selection of those pieces to be read and considered. Both of these approaches place a premium on the quality. Institutions should choose evaluation methods that emphasize quality rather than quantity, and authors should provide their readers with truly meaty slices of their baloney.

Reporting of publications on resumes

It is particularly important for geographers to accurately communicate to readers of their resumes the exact status of their publications at the date shown on the resume. Citations of published articles include the volume and page numbers, and without such information articles should not be considered "published." Articles should not be listed as "in press" or "accepted" unless the author has a written communication from the editor that contains specific and direct language that the paper has been accepted. A letter indicating that the paper is acceptable upon revision is not enough, because it means that the author has not yet completed satisfactory revisions, and the editor has not yet evaluated the revisions, or actually accepted the final paper. If authors insist on listing in their resumes papers not yet accepted, they should include a parenthetical statement indicating the precise status of the paper at the date of the resume, such as "submitted to the editor for review," "under revision," or "in preparation." Jointly authored articles should be reported with the order of the authors' names as they appear on the original publication.

False or misleading reporting of the status of publications on resumes is particularly damaging, because it gives an untrue picture of the author's record, and because it establishes a pattern of stretching the true nature of the situation. The practice is surprisingly common. It is already difficult for committees reviewing job applicants, candidates for promotion, or grant applicants because of the wide range of specialties and publishing ethics we deal with in geography, a discipline populated by social scientists, natural scientists, and humanists. Precise reporting of publication status is also a question of simple personal honesty and self-confidence; it is not a situation in which to make false claims which may embarrassingly not prove true in the long run. Once discovered, imprecise reporting naturally leads to suspicion about other aspects of the resume, and a general erosion of confidence. Like self-plagiarism and slicing the baloney thin, imprecise reporting of publications does us all a disservice, and whatever the perceived rewards of such actions, they just aren't worth it.

The issues of self-plagiarism, slicing the baloney thin, and inaccurate reporting of publications on resumes are important subjects of discussion between mentors and those with whom they work, as well as among professional colleagues generally. Formal training in ethics is becoming more common in private industry, and it should be part of the standard curriculum in our graduate education.

Susan Cutter: *Big Brother's New Handheld*
May 2001

Privacy is rapidly becoming the issue of the 21st century. While many retain visions of George Orwell's *1984*, the reality of the information age makes his novel tame by comparison. The right to privacy is guaranteed in this country. Presumably embedded in this freedom is the right to hide. Current technology eliminates this fundamental right. Further, the geographical community is inadvertently contributing ideas, technology, software, training, and expertise that can be used to locate and thus monitor individual behavior remotely—a clear violation of privacy protections.

The availability and use of highly precise tracking systems necessitates ethical consideration by the discipline. What moral and social responsibilities govern the use of GIS? How can the privacy of individuals be assured when utilizing such systems (remote sensing, GIS, GPS) for teaching and research? Where is the line between such innocent use of the technology as tracking hazardous material trucks, and more sinister applications such as inserting microchips under the skin to track the movement of employees, spouses, or children? This may seem like a world of action movies and spy novels, but the technology already exists.

A recent front-page story in *The New York Times* described the burgeoning growth of wireless tracking systems. These locational technologies are one of the fastest growing segments of the wireless communications industry, with annual revenues of $600 million annually now, forecast to reach $5 billion annually in three years. Installation of GPS or geocoded tracking systems in vehicles (including rental cars), hand-held computers, cell phones, and even bracelets and watchbands is already happening. James Bond is not the only person who can acquire such gadgets.

There are positive aspects to the wireless location-based services: enhanced 911 service, for example, allows a caller's address to be pinpointed when he or she dials the emergency network. In fact, the federal government has been one of the leading proponents of such technology through its advocacy of improved geographic precision in emergency 911 calls. In August 1999, the US Federal Communications Commission required all wireless providers to include tracking devices in cell phones so the location of the beginning of the call and the end of the call were available (see Center for Democracy and Technology at *http://www.cdt.org*). The use of wireless GPS in real-time navigation is positive as well, although I worry that drivers will lose their ability to read maps, preferring instead to "listen" to navigation directions from a receiver installed in the vehicle's dashboard.

From my perspective, the potential dangers of personal tracking devices outweigh their social benefits. I am especially concerned about the rights of women and children, and their ability to resist having the technology imposed on them. Scientists have developed a locator chip that can be embedded under a person's skin. Instead of having your teenage children call home to check in, you simply dial them up on the computer to verify their whereabouts. While this may allay parents' fears, what about the rights of the children? Wouldn't implantation of these devices also facilitate child abduction by computer literate, non-custodial parents? The use of personal tracking systems is a considerable concern, given that GPS locations are often off by several meters or more—an error that could place a child in a forbidden movie theater rather than across the street where she really was. Or what about the ability to track and monitor child workers in the world's sweatshops and send electric shocks to those who stray from the assembly line?

Women's rights and freedoms may also be at risk. There may be more resistance to embedding locator chips in women, but GPS tracking from cell phones or devices placed in wristwatches is a more subtle probability. Women's daily activity patterns could be monitored either by possessive spouses, potential stalkers, or disapproving employers or governments. A woman's ability to lead an independent life would be compromised by constant surveillance. Because the technology is male-dominated, I suspect that the gender reversal would not be prevalent. Personal freedoms could become the first casualty of a new locational e-slavery. A case in point is found on the Digital Angel website (*http://www.digitalangel.net*)[1]. When logging on, the visitor is taken through three cartoon animations—Grandpa having a heart attack at home, Jane having car trouble, and Spot (the dog, of course) wandering away. In each case, the protagonist is rescued via the use of GPS, depicted in the cartoons as a white angel. It all seems innocent, but is it? [*Editors' note: At the time of printing (2004), the trademark Digital Angel ® was being used to market tracking systems for pets, not humans. However, Digital Angel's parent company, Applied Digital Solutions, was marketing a similar implant for humans under the name VeriTech ®. See* http://www.adsx.com/content/index.htm *for more information.*]

In the right hands and under the right conditions, these locator systems provide some social benefits. But who is to say which conditions and applications are right? I'm not convinced that these decisions should be left to wireless carriers or their customers. Nor am I convinced that government can protect the privacy of citizens. Surely there are some nations in the world that would not embrace these technologies. Consequently, the professional community should address these issues in classrooms, workplaces, and research. We need to discuss and debate the ethnics and function of personalized locational systems and formulate appropriate guidelines for their use. What may appear initially as a benign technology or application, may have untoward and unanticipated consequences that deserve consideration. If we ignore these issues, aren't we turning a blind eye to the e-slavery of elderly, children, and women, now and in the future?

For more information see Simon Romero, "Locating Devices Gain in Popularity but Raise Concerns," *The New York Times*, March 4, 2001: A1; and Jerry Dobson, "The 'G' in 'GIS'": What are the Ethical Limits of GIS?" *GEOWorld*, May 2000 (*http://www.geoplace.com/gw/2000/0500/0500g. asp*).

✳ ✳ ✳

Chapter 13

HUMOROUS MUSINGS

❋ ❋ ❋

To illustrate that not all Presidents took a serious tone in their writings on important matters facing the profession, we include a sampling of the more humorous columns that were written. There is a tinge of truth in all of them. Often, these columns struck a chord with readers and prompted the most response from the membership!

❋ ❋ ❋

BOB KATES
November 1993

Boulder, Washington, Milwaukee . . . so begins this President's transhumance. The President grazes on our grassroots and tries to leave something in return.

What is left differs—by regional tradition, by setting, by program, by president—but there is always one staple for opening greetings, introductory welcomes, plenary keynotes, or after lunch and dinner remarks; jokes.

So Bob do you know any geographical jokes? No, no, and almost no one else knows any either. At least that was the result of my sampling at the Great Plains–Rocky Mountains regional meeting. In a room almost filled with 150-200 geographers, in lieu of my own joke ignorance, I asked: Does anyone here know any geographical jokes? Only one hand is raised in the entire audience, and she is a Boulder alumna of the class of 1933. She knows one joke and has had 60 years in which to learn it or create it!

So why don't we know more jokes? It can't be the aridity of the plains or the chill of the mountains. The AAG Council, representative of the whole country, fares no better a week later. They don't know any jokes. At least no real geographic jokes, not just the use of "geography" as a generic for place.

Real geographic jokes must spring from the essence of our disciplinary understanding—not from recycling or substituting geographers for Frenchmen, lawyers, Yeltsin, or even economists. I fear there is a heaviness to our discipline (despite an occasional chuckle in these pages from Khaldun or Ptolemy), and perhaps to all disciplines. Know any really good academic jokes?

There are of course exceptions. We all know the stories of great entertaining teachers. And increasingly there are a few attempts to turn our professional practice into celebration and festival. We tried a bit with the rededication of the AAG Office in Washington on 19 September. Of course $435,000 is no joke, but there were balloons, a purple dinosaur, resilient sponge ball earths for all, and best of all, kids. Don Dahmann led the Children's Neighborhood Geographical Expedition. And to commemorate his remarkable photo—essays in past issues of the *Newsletter*, Ron Abler received the Cracked Ceiling Award for "mastery of the arts of the built environment, perseverance in the navigations of the regulatory landscape, and creative pursuit of budgetary balance."

Our effort to lighten up takes second fiddle to the greatest geographic celebration in the world, the St. Dié des Vosges Festival International de Géographie, now celebrating its fourth year almost as I write these words. This small town's claim to fame, besides the most entrepreneurial of geographic mayors, is as "the town that put America on the map." For here in 1507, the name "America" first appeared on a map employing Vespucci's new geography and according to legend, feminized to match Asia and Africa. There is a serious side to the festival, lectures, and panel discussions, and even awards to leading geographers. But for thousands that throng the town each year, it is a veritable celebration of books, maps, globes, displays, new technologies and views of the world in ways that excite the mind and tickle the funny bone, especially of children.

So as we struggle to putgeography on the mental maps of necessary knowledge, can we also lighten up a bit? To do my part, I hereby announce the FIRST (and probably the last) PRESIDENTIAL GEOGRAPHIC JOKE COMPETITION! Winners will be my guests at the San Francisco banquet (and guess whose jokes I will tell that night). And did you hear the one about the geographer and the...

RON ABLER: *Five Steps to Oblivion, II*
January 2000

It's been twelve years and six months since my prose appeared in this space. Reg Golledge wanted a month off and invited me to fill in for him. My musings about a topic were interrupted by a call from a geography department that had just been placed on the chopping block by its university curriculum committee. That gloomy news called to mind a little essay published in 1993, which seems worth a second outing:

"It has come to my attention that some college and university faculty remain untutored in the techniques that ensure a program's demise. These parlous times in academia would seem to guarantee that any program harboring the slightest death wish would soon be accorded martyrdom, yet some college and university administrators remain stubbornly convinced of the intellectual value of certain programs, despite their lackluster performance.

Herewith, therefore, a sure-fire prescription for overcoming administrative resistance to closing a program. It will work even in the most refractory cases.

1. Elect and re-elect a weak chair who cannot lead the faculty to consensus on an appropriate program mission, and who cannot articulate the program's contributions to campus colleagues and to administrators who allocate resources. (This is a critical step; success here virtually assures that several subsequent steps will follow automatically.)

2. Lose several of your most productive colleagues, upon whom much of your program's intra- and extramural reputation depends. Do not replace them with individuals of comparable accomplishment and prestige. University administrators will never compare the program's past glory with its present status when they decide how to allocate scarce resources.

3. Eschew undergraduate education and majors; they are clearly beneath your program's dignity. The preoccupation of university officers with statistics on undergraduate instruction is typical administrative bean

counting. If those pesky undergraduates persist in trying to enroll in your program, assign your largest courses to your worst instructors or to marginal graduate students who need to learn how to teach. Make sure advising is slipshod and callous.

4. Forego participation in campus governance, which is well known to be a complete waste of time. A quality program consisting of thoughtful scholars will never need friends in key campus positions. Curriculum committees are especially noxious; avoid them like the plague.

5. Glory in bitter ideological and personal vendettas. Never ever fail to denounce your opponent(s) to your dean and provost. They have infinite patience with internal bickering and they will never conclude that both sides of the dispute have correctly assessed their adversaries.

Any three of these steps will normally place a program on the endangered list. Four should suffice even where administrators really value your subject. All five will normally overcome any reservations held by even the most ardently committed dean or provost. Yet there are, unfortunately, instances in which extreme measures are necessary.

6. When all else fails, war against your dean. There is no chance that he or she will prevail against a mighty program like yours.

Since 1993, a number of programs seem to have missed the irony in this tongue-in-cheek prescription—a moment's reflection on the history of any program that has been terminated, merged, or endangered in the last seven years will reveal one or several of these processes at work. Through my immersion in AAG affairs since 1984, I've learned more than I ever wanted to know about failing and failed departments, and it seems to me that the caveats encapsulated in the five steps plus one continue to offer good advice for the future.

Four additional points seem worth stressing. First, threatened programs are, almost by definition, beyond salvation. By the time university committees and administrators talk openly about eliminating or merging a program neither frantic action by faculty and alumni nor intervention by professional societies such as the AAG can do much good. Universities and their administrators view gratuitous advice and opinions from outsiders as unwarranted interference in their internal affairs. In human terms, it's well nigh impossible for a committee or administrator to retract a proposal for merger or termination once it has become public. Moreover, such recommendations usually damage the local credibility of the program irreparably.

Second, it follows that the time for marginal or mediocre programs to take remedial action arises well before (years before) they become threatened. If one or more of the six steps to oblivion sounds uncomfortably familiar for your program, get to work immediately. Seek external assistance and advice when presented with a tightly reasoned case that your program is doing well but could do more for its college and university with the benefit of an external evaluation. Most administrators will provide the minimal financial support needed for an external evaluation, and if your institution cannot provide the means, the AAG will find a way to put competent advisors on your campus. Involving key administrators in the review is critical; because they have asked for it, they will be morally bound to accord the external advice some consideration.

Third, in fifteen years of trying to rescue troubled and threatened programs, I have met but one university administrator who may have acted out of malice toward geography and geographers per se. That's the exception that proves the rule that no sane dean or provost will harm or hassle a program that contributes effectively to his or her institution. Conduct your program in ways that give your dean bragging rights among his peers and your provost bragging rights among hers, and you'll never fear oblivion. Sadly, almost all geography's wounds have been self-inflicted.

Fourth, bear in mind that merger or termination almost always means oblivion. There was a time when academic institutions could afford to clean house and rebuild troubled programs. Rehabilitating dysfunctional or unproductive programs is now more difficult given current financial constraints. Moreover, geography is not among the disciplines administrators view as essential to campus life. A university administration may believe it has no choice but to rehabilitate an ailing or failing history program. Geography will likely be viewed as expendable, and degrees and programs lost now are almost certainly gone forever.

Dark ruminations for a new year, decade, century, and maybe even millennium (depending on your numerical theology)? No question. Helpful considerations for such occasions? I think so. Geography prospers these days, as is evident in the record number of positions advertised in this and the previous issue of the *AAG Newsletter*. Yet components of the enterprise continue to self-destruct. Let us therefore adopt Pogo's famous dictum as the watchword for the beginnings 2000 offers: "We have met the enemy and it is us."

Susan Cutter: *Steps to Insure an Unsuccessful Personnel Search*
November 2000

This is the time of year when the number of advertisements in *Jobs in Geography* increases, as academic departments secure permission to hire new faculty for the next school year. The recruitment of new faculty is one of the most important decisions academic departments make (along with tenure and promotion) and the decision-making is often contentious, even in the most collegial of units. Given the significance of hiring (it doesn't matter at what level), I have been amazed how clumsily we often handle the search process and how dumbfounded we are when our recruitment efforts fail.

I therefore offer my insights into the search process based on my experience as chair of numerous search committees and as recipient of the horror stories told by my graduate students as they return from interviews. IF your goal is to have a less than successful search, just follow the steps outlined below.

1. Request reference letters from all candidates as part of the application process. This tactic will not only bury you in paperwork, but will enable you to get the least informative letters possible. Referees will simply change the name and address of the generic word-processed letter, and not alter its content for your specific position. If you're lucky the letter will be current and not recycled from last year's job search.

2. Develop an advertisement for everything, so that no one person fits the description. If you don't want to hire in any given year describe a position for which very few people can qualify, using a number of specialties that don't really fit most candidate's profiles. A remote sensing human geographer with Latin American regional experience who can teach advance climatology, social theory, and spatial statistics is one example. If your want a shallow pool of candidates from which to choose, write a shallow ad.

3. Make sure the search committee has a majority of soon-to-be retired faculty members. Those established tenured faculty who lament the direction of contemporary geography and who also have firm and vocal agendas about replacing their subfields (and thus themselves) are

best. Keep in mind that these colleagues will not have to live with the hiring decision that is made.

4. Never send an acknowledgement letter and never let the candidate know where she stands in the hiring process. This is a surefire way to get e-mails and phone calls from anxious candidates who have no idea whether materials were received, or the status of their applications. If you want to foster a couldn't-care-less culture and inhospitable work environment, this will help greatly.

5. Ask the candidate to stay over a Saturday night to reduce expenses and then fail to handle local arrangements. This entails having the candidate find his own way to the hotel, leaving him alone Saturday night and all day Sunday, and then leave the candidate to find his own way to the department for the first appointment on Monday. This will help ensure a nice weekend free of obligations for members of the department, but it certainly doesn't impress your candidate, especially if you are trying to convince him that your university is the perfect place to start his career.

6. Put your most cantankerous faculty member in charge of hosting the candidate during an on campus visit. This is almost a given for any failed search. The faculty member will air all the department's dirty laundry, probably offend the candidate at some point during the visit, and inform the candidate that the senior faculty never promote or tenure anyone. What a way to build the image and reputation of the department.

7. Set the itinerary for the campus visit with meetings scheduled for each faculty at time when individual faculty are unavailable. This will ensure that the candidate knows how busy faculty are and how inconvenient it is for them to talk to anyone, including the candidate. When rescheduling the appointment, be sure that your colleagues intimidate the candidate during the one-on-one conversation with esoteric and off-the-wall questions. This helps render the interviewee speechless and dazed for the rest of the visit, and enables faculty to opine that the candidate knows nothing about the discipline or their specialties.

8. Make sure that the room for the talk has no audio-visual equipment, or if it does, make sure it isn't working properly. The idea here is to make the presenter as nervous as possible. Faulty overhead projectors, jammed slide projectors, or inoperable PowerPoint ® presentations

should ensure this. There's no need to test the equipment before hand, since the audience really does like to see the candidates squirm.

9. Once the interview is over, take months to reimburse the candidate's expenses. Not only will this help the candidate to understand the bureaucracy of your university, it will keep your departmental budget in the black a little longer.

10. Last, but certainly not least, never let the unsuccessful candidates who visited your campus know they weren't hired. This will save you considerable correspondence. You won't have to send rejection letters to long-listed or short-listed candidates. It will also enable you to string a candidate along for as long as possible and feel a sense of power over someone's future.

If you follow this ten-step program, I assure you that your pool of candidates and the ultimate success of your search will be just what you wanted—incomplete, inadequate, and unacceptable.

Unfortunately, all of these examples have a ring of truth and happened in some form to many of the brightest and most energetic colleagues in our field. Do your experiences fall into one of these categories? I would welcome additions to my collection of rules on how not to run a job search.

✳ ✳ ✳

Chapter 14

CONCLUDING THOUGHTS

✳ ✳ ✳

M. Duane Nellis, Janice Monk, and Susan Cutter

Just as the AAG Presidential columns provide a brief snapshot to the current issues of the day as well as long-term trends in the discipline, our concluding thoughts represent a brief synopsis of the value of creating a book of such musings. We also offer some of the responses Presidents' received from members as testimony to the lively engagement of geographers with professional issues. Presidential columns have become one of the most read communications from a geographer to our membership, but the *AAG Newsletters* are often read and discarded, much like one would do with a daily newspaper. Through archiving of selected past presidential columns, we have attempted to pick the most relevant (at least from our perspective and those of the past presidents) for long-term recording and reflection.

The columns reveal ways in which the discipline of geography has evolved over more than three decades. New themes have surfaced in these musings (for example, the role of geospatial technology). But there are also continuities, such as issues about membership and concerns about the roles and qualities of publications and the most appropriate formats for annual meetings. The columns reveal that the discipline has gone through times of outside challenges and internal discourse and debate related to new directions in research and the extent to which we reach diverse audiences and influence public understanding and policies. They also pose important challenges for our future, especially those of diversifying our membership and exploring how the Association can be more inclusive in serving those who work in a variety of diverse professional settings in addition to those in academia.

Though Presidents have used the columns to advance their own concerns and perspectives, most have also regularly included references to the work of an array of geographers over the course of their year's columns and invited readers to send their reactions to the positions expressed, to ask questions, and to supplement the information included in the columns. And many readers certainly took up the offer to respond. While all Presidents routinely engaged in private correspondence with members of the Association, several Presidents, in fact, devoted whole columns, or major portions thereof, to readers' responses. While we have not included full versions of these responses in this volume, we want to give some flavor of them and have incorporated one such column as an example.

Among columns generating the most dramatic and divided responses were those related to changing the Association's publications and priorities. Terry Jordan, for example, was inundated with letters in response to his columns (Chapter 11) about placing the Association's highest priorities on scholarly publication. Some supported his position, others disagreed. Fred Kniffen, for example, wrote: "I heartily agree with your stand in support of better funding for our publications. If you don't accomplish another thing, your term will be a rip-roaring success." Conversely, Susan Brooker Gross saw him as creating "an unnecessarily adversarial relationship between scholarship and educational ground work. Life is not so simple: scholarship, in turn, depends upon all those other activities." Will Graf, who sought the membership's views on the philosophy and structure of the AAG's publications, likewise received responses from a wide array of geographers ("The Readers Write Back", 33 (12), 1998). They generally expressed support for *The Professional Geographer* but were much more divided about the nature of the *Annals*, questioning its philosophical underpinnings, balance of the areas of research, format, and lack of color. Some suggested combining the two journals. The eventual resolution took these readers' comments seriously, dividing the *Annals* into thematic sections to assure inclusion of major sub-areas of the discipline and introducing some color printing.

Judy Olson, in her last column, entitled "Feedback and Farewell" (June, 1996), acknowledged resources and ideas that geographers had sent her to complement and expand on the information she had offered (see Chapters 5 and 7) on enhancing undergraduate education, uses of cartography, strengthening regional meetings, and on administration within universities. To reveal how her readers engaged, we reproduce here the entire column.

※ ※ ※

Judy Olson: *Feedback and Farewell*
June 1996
Over my year as President of AAG, I have appreciated those who have commented on these columns. Some of the responses have already appeared; I would like to devote this last column to some of the others.

In response to the July 1995 column concerning the cooperative efforts between the University of Kansas and Haskell Indian Nations University, David Niddrie, Professor Emeritus at the University of Florida, sent me a piece from his alumni publication (University of Natal, South Africa). Written by Robert Preston-Whyte and Barry Angus, the article states:

> *"[T]he lecture theatre . . . is where the problems produced by decades of unequal education must be confronted. This is where solutions must be found to cope with the reality of varying competence in language and numerical skills. This is where the success of the transformation will be translated into competent graduates and satisfied employers."*

The article goes on to describe an overhauling of the Environmental Science course to incorporate opportunities for student skill building, availability of trained tutors, quality control in instruction, and meaningful course evaluation. The multi-pronged approach resulted in measurably better student performance and radically fewer failures without compromising standards, and after two years "it was... no longer possible to identify failing students along ethnic [or correlated] lines." The importance of the effort is self-evident.

When I was preparing the September 1995 "Evil (?) Administrators" columns, an inquiry to Ron Abler resulted in his sending me a series of editorials by John Adams, who has been chair of the Faculty Consultative Committee at the University of Minnesota. The titles of the editorials include: "Faculty must take proactive approach to U issues," "Leading well without over-managing," and "Understanding Responsibility Center Management." My column went off on a rather different angle, but John's materials may well be of interest to readers and you may contact him for copies.

I was always ready to hear suggestions for column topics and a colleague who shall remain anonymous suggested "[something] all geographers...would do well to think about...[I]f we are in a communications age, and everyone is communicating graphically these days...where are the statistical maps from the Census Bureau?...[W]hy just tabular presentations?" I will not try to answer the question, just point it out. I have heard it raised by various colleagues since the early 1980s. I suppose one argument is that modern software has made all of us into mappers and we can make our own maps. It is an "answer" that raises a lot of questions. The suggested topic would have been good material for a column, but perhaps there will be another forum or another columnist to take up the issue.

In the correspondence related to the October 1995 column on "The World Wide Web," Ken Foote at the University of Texas provided some data on usage of their materials. " . . . [L]ast week we 'served' 28,654 files, almost 260 megabytes of information . . . Our modules on technical issues like GPS, map projections, coordinate systems, and datums are receiving over 3,000 external file calls a week." The usage is probably up from those now—outdated figures. They are an indication that Web pages are at least being found and that communicative effectiveness via the Web has considerable potential.

Speaking of the Web, we used it as a vehicle for answering questions posed by the audience at the Presidential Plenary Session ("Is GIS Killing Cartography?") in Charlotte this year. The results are intended for the audience that was actually there, but if anyone wants a look, you will find them at *http://www.ssc.msu.edu/~geo/AAGanswers.html* [*Editors' note: At the time of printing (2004), this link is no longer available.*]

In the January 1996 column, "Regional Division Meetings," Cathleen McAnneny from the University of Maine at Farmington pointed out an important omission—the importance of these meetings to undergraduates. Dr. McAnneny knows of what she speaks. At the New England/St. Lawrence Valley Division (NESTVAL), she and her colleagues had a wonderful and sizable group of undergraduates, one of whom won the student paper award. As she points out, regional meetings are an opportunity for undergraduates to interact with other faculty members, develop ties with peers from other institutions, and explore questions about graduate schools as well as give papers and participate in the Geography Bowl.

The latest correspondence (which started on April 1!) is in response to the April 1996 column "Annual Meeting Sessions," in which I raised the question of alternatives to the usual paper-session format. The alternatives all center on "active learning." Nancey Green Leigh, Vice President of the American Collegiate Schools of Planning said thanks for "food for thought on how our conference sessions can be more effectively organized." Susan Hardwick summarized her response as "Yes Yes Yes." Robert Dilley was favorably inclined toward more poster sessions but pointed out that there may be a need for convincing those who authorize travel funding that posters are equivalent to oral papers. Let me go on record here as fully supporting this equality. The presence of the presenter is crucial to poster presentations, and we expect the communicative value of posters to be on par with oral papers. Posters are NOT rejected oral papers; they are a different form of presenting equally worthy material.

Adena Schutzberg, from ESRI, indicated that the possibility of new session formats was causing her to perk up her ears and consider attending AAG again. She pointed out that such formats are more common at GIS conferences. She also raises the problem of presenter preparation for alternative formats and the need for guidelines, experience, and training if these formats are to work.

John Fieser, Economic Development Administration, argues that because oral papers are launching pads for publications, alternative formats are not "serious competitors," and that improvement with the traditional format is what is really needed. One of his statements is particularly important: "Innovation could well turn out to be as sloppy in its execution as paper sessions now are, gaining us nothing." He is absolutely right on that point. The idea of "active learning" can be applied in many ways, however, from improvement of our "regular" papers to the use of completely different session formats. Wherever we are on the scale of change, I hope we all have quality and improvement as our motivations, not change for the sake of change.

Thanks to all who have commented over the past year! I have enjoyed writing these columns far more than anticipated. They have been more of a learning experience than a chore. Your responses were an important reason for the rewarding experience.

❋ ❋ ❋

The AAG and geography have a long and dignified history, of which these columns reflect one small part. We hope through this collection of "presidential musings" we can learn from the past and create an even stronger future for this essential discipline. Geography thrives by giving young minds—in this case the "new milennials"—the opportunity to ask new questions and to create new geographies as we attempt to explain order or patterns in occupied space on the face of the earth. The Association has an important part to play in fostering such work. It is a challenge not only for the leadership, but for the membership. As George Demko wrote in his final column (June 1987):

> *"Most important . . . is to encourage all of you to get more involved in the Association . . . Do you know how your money is spent? Are you aware of what programs the Association promotes and how support is given for them? What does the Association do for you? You can get more involved by using your councilors, officers, and the central office staff. Send in ideas, volunteer, complain when responses are not forthcoming. Although it sounds like a cliché: it is your Association, supported by your money—be part of it!"*

Future presidents, no doubt, will surely have perspectives to offer us on the hot topics of the day and future directions. We look forward to reading these columns as they chart the geography of the 21st century. We give the last word to Bob Kates, who shared some of his own personal reflections in his last column (June 1994).

> *"But the most important way to strengthen our external connections is to do so in our daily pursuits. When geographers everywhere, in government, in the university, in the private sector, are seen as the connectors, as bridge builders, as integrators, the folks who take the initiative to bring folks together, then will our discipline be solidly anchored by an extraordinary root system capable of withstanding the hazards of our times: the storms of restructured*

institutions, the droughts of shrinking budgets, and the faultlines of changing intellectual interests. As this year of being President concludes, I share a final discovery. Much to my surprise it has been a year of quiet satisfaction and real pleasure: satisfaction for the good health of our profession at a time when everyone is called upon to do more with less, and pleasure for a rewarding personal voyage of geographic rediscovery."

* * *

Chapter 15

BIOGRAPHICAL SKETCHES

※ ※ ※

RONALD F. ABLER is Secretary General and Treasurer of the International Geographical Union (IGU). He was Executive Director of the Association of American Geographers (AAG) from 1989 through 2002. From 1984 to 1988, Abler was Director of the Geography and Regional Science Program at the National Science Foundation, where he expanded funding for physical geography and coordinated the establishment of the National Center for Geographic Information and Analysis (NCGIA). Abler's research explores the ways different societies have used intercommunications technologies at different times and places. He has written numerous research articles and is co-author or editor of several books. Most recently he edited *Global Change and Local Places: Estimating, Understanding, and Reducing Greenhouse Gases.* He has received the Centenary Medal of the Royal Scottish Geographical Society (1990), Association of American Geographers Honors (1995), the Victoria Medal of the Royal Geographical Society/Institute of British Geographers (1996), and the Samuel Finley Breese Morse Medal of the American Geographical Society (2004). Abler was President of the AAG in 1985–86.

JOHN S. ADAMS is Professor of Geography, Planning and Public Affairs, and Chair of the Department of Geography, University of Minnesota, where he holds the Fesler-Lambert Chair in Urban and Regional Affairs. He received his BA in economics and mathematics from the University of St. Thomas in St. Paul in 1960, and his PhD in economic geography and the University of Minnesota in 1966. He taught at Pennsylvania State University, and was visiting professor at the University of Washington, and at the US Military Academy at West Point. His research focuses on the American city, regional economic development, intra-urban migration, housing markets, urban transportation and urban development in the USA, and the states of the former Soviet Union. His courses and seminars address methods for analyzing population and housing, land use and transportation, and the metropolitan economy; the Twin Cities of Minneapolis-St. Paul; and Russia and environs. He currently works with the Minnesota Department of Transportation on land use and transportation issues facing the state's major urban areas, and is co-principal investigator of the National Science Foundation-sponsored National Historical Geographic Information Study. He was awarded AAG Honors in 1988. Adams was President of the AAG in 1982–83.

BRIAN J.L. BERRY is Lloyd Viel Berkner Regental Professor and Professor of Political Economy at the University of Texas at Dallas (UTD). He received his BS (Economics) at University College, London (1955), and MA/PhD (Geography) at the University of Washington (1956–1958). He was a faculty member at the University of Chicago (1958–76), where he chaired the geography department and directed the Center for Urban Studies; at Harvard (1976–81) where he chaired the doctoral program in urban planning and directed the Laboratory for Computer Graphics and Spatial Analysis; at Carnegie-Mellon (1986–91) where he was Dean of the Heinz School of Public Policy and Management. He joined UTD in 1986 and was founding director of the Bruton Center for Development Studies. The author of more than 500 books and articles, he is best known for his work in spatial analysis, location theory, urban and regional development, and on long waves in economic, political, and social change. His work attempts to bridge theory and practice via applied research and advisory work with policy makers. He is a member of the National Academy of Sciences, the first geographer to be elected to its Council, a Fellow of the British Academy and American Academy of Arts and Sciences, and received the Victoria Medal from the Royal Geographical Society (1988). He was President of the AAG in 1978–79.

STEPHEN S. BIRDSALL is Professor of Geography at the University of North Carolina in Chapel Hill. He received his BA (1962) in Earth Science from Antioch College and his MA (1964) and PhD (1968) in Geography from Michigan State University. He has been on the faculty at the University of North Carolina at Chapel Hill since 1967. He served as Interim Dean and then Dean of the College of Arts and Sciences at UNC–Chapel Hill from 1991-97, after which he returned to the joys of scholarship and teaching as a regular faculty member in the Department of Geography. He is currently engaged in research on natural and memorial landscapes. His service to the AAG includes a role in the Southeastern Division as an officer of the Division or as a member of the Steering and Executive Committees for 11 years during 1972–87. He served on AAG Council for three three-year terms between 1984 and 1996, including five years as Treasurer. Birdsall was President of the AAG in 1994–95.

JOHN R. BORCHERT (1918–2001) was a Regents Professor of Geography at the University of Minnesota. From a wartime career as a meteorologist he earned his PhD in geography at the University of Wisconsin, Madison in 1949, and then went to the University of Minnesota where he began his teaching career (1949–1989). His wide research and consultancy work spanned issues in climate, natural resources (especially water resources), land development and settlement in the US, urban development and the evolution of metropolitan centers, economic regions, highway development, and public land policies. Among his many publications is the prize-winning book, *America's Northern Heartland*. Borchert brought his geography to many public roles, serving as a consultant to the Legislative Commission on Minnesota Resources, the Twin Cities Metropolitan Planning Commission, the Board of Minnesota Pollution Control Agency, and others. He was founding Director of the Center for Urban and Regional Affairs at the University of Minnesota. He was a member of the National Academy of Sciences, Chairman of the Earth Sciences Division of the National Research Council, Member of the National Research Council Transportation Research Board, and Chair of the US National Committee of the International Geographical Union. Borchert was President of the AAG in 1968–69.

LAWRENCE A. BROWN is a Distinguished University Professor at Ohio State University where he served as Chair of the Department of Geography from 1995–2003. His interest in geographic study was piqued during two college-era travels—working his way across the United States and driving the Pan American Highway. His research has focused particularly on innovation diffusion; migration in both national and international settings; Third World and frontier settlement development drawing on the experiences of Latin America; regional socio-economic change in US settings related to the Fordist/Post-Fordist transition; and most recently, racial/ethnic dynamics of US cities. Brown's research record is strongly linked to advising and mentoring students, many of whom have become prominent professionals. His best-known publications include the books *Innovation Diffusion: A New Perspective* (Methuen, 1981) and *Place, Migration, and Development in the Third World: An Alternative View, with Particular Reference to Population Movements, Labor Market Experiences, and Regional Change in Latin America*. Recognitions include an Honors Award from the Association of American Geographers (1983), Guggenheim Fellowship (1985), and Distinguished Mentor Award of the National Council for Geographic Education (1996). Brown was President of the AAG in 1996–97.

SAUL B. COHEN, a Boston native, received his BA, MA, and PhD degrees in Geography from Harvard, after serving in World War II. He has been Professor of Geography at Boston University, Director of Clark University's School of Geography, University Professor at Hunter College, held visiting professorships at the US Naval War College, Wellesley College, Yale, and the Hebrew University, and is the recipient of five honorary degrees. He was President of Queens College, CUNY from 1978 to 1985. In 1965–66 he served as visiting Executive Director of the AAG. He has also been an Association Councilor and served on numerous committees, spearheading the establishment of the Commission on Geography and Afro-America, Committee for Recruiting Talented Minority Students (COMGA), and the Commission on College Geography. He has been geography consultant to the National Science Foundation and chaired the National Academy of Sciences–National Research Council Committee on Geography. Cohen's interest in educational reform began with membership in national educational committees in the 1960s and 1970s, extending to membership in the New York State Board of Regents (1983 to present), where he has chaired the Elementary and Secondary and the Higher Education committees. He is author/editor of thirteen volumes and over one hundred articles dealing with political and economic geography and the Middle East. His most recent book is *Geopolitics of the World System* (2003). Cohen was President of the AAG in 1989–90.

SUSAN L. CUTTER is Carolina Distinguished Professor of Geography at the University of South Carolina, where she also directs the Hazards Research Lab and served as chair (1993–2000). She received her BA from California State University, Hayward (1973) and MA (1974) and PhD (1976) from the University of Chicago. Cutter has served on the faculties at the University of Washington and Rutgers University, the latter where she also chaired the department. Her research interests are in environmental risks and hazards, vulnerability science, and spatial decision making in response to environmental threats. She is the co-founding editor of the interdisciplinary journal, *Environmental Hazards*. Cutter has authored or edited twelve books, more than 100 scientific papers, book chapters and reports, more than 35 successful grant proposals, and more than 100 public presentations. Her most well known books include *The Geographical Dimensions of Terrorism* (Routledge), *American Hazardscapes* (Joseph Henry/NAS Press), and *Living with Risk* (Edward Arnold). A strong advocate of the use of geography in public policy, she has served on many national advisory boards including those of the National Research Council, NASA, and the H. John Heinz III Center for Science, Economics and the Environment. She is an elected Fellow of the AAAS. Cutter was President of the AAG in 2000-2001.

GEORGE J. DEMKO is Professor Emeritus of Geography at Dartmouth College and a faculty member at Charles University in Prague. He received his PhD from The Pennsylvania State University and has been a visiting professor at a number of institutions including Moscow State University and Swarthmore College. A specialist on geopolitics and population geography, he has written 17 books and more than 90 articles. He is also a specialist on international crime fiction and writes a regular column on mysteries in foreign countries. Demko has been Geographer at the US Department of State. He received the Gold Medal of Charles University for a lifetime of contributions to geographical knowledge and the promotion of international intellectual cooperation. Demko is a Fellow of the Russian and Slovak Geographical Societies and the recipient of many awards and honorary degrees. Demko was President of the AAG in 1986–87.

PATRICIA GOBER is Professor of Geography at Arizona State University where she served as departmental chair from 1984–1991. She received a PhD in Geography from Ohio State University in 1975. Her research centers on issues of migration, immigrant communities, retirement communities, and environmental change in metropolitan Phoenix. Her publications include a 1993 Population Reference Bureau monograph, *Americans on the Move*, more than 50 articles in geography, demography, and environmental studies journals, and *Human Geography in Action*, a textbook and CD-ROM for use in introductory human geography courses. She is currently Vice Chair of the Board of Trustees of the Population Reference Bureau and Chair of the College Board's Advanced Placement Human Geography Committee. She served on the National Institute of Health's Social Science Methods, Epidemiology, and Nursing Study Section (1998-2003), was a member of the Science Advisory Board of the National Oceanic and Atmospheric Administration (1998–2001), and received an honorary doctorate of science from Carthage College, Kenosha Wisconsin in 1998. Gober was President of the AAG in 1997–1998.

REGINALD G. GOLLEDGE is Professor of Geography at the University of California, Santa Barbara. He earned his BA and MA at the University of New England (Australia) and his PhD at the University of Iowa. Legally blind for the past 20 years, he has interests in behavioral geography including spatial cognition, cognitive mapping, individual decision-making, household activity patterns, gender issues in spatial cognition, and the acquisition and use of spatial knowledge across the life span. Recent research has included work on adventitious and congenitally blind persons. Golledge's awards include Honors from the Association of American Geographers (1981); a Guggenheim Fellowship (1987-88), and the Australia–International Medal from the Institute of Australian Geographers (1998). Golledge was President of the AAG in 1990–91.

WILLIAM L. GRAF is Educational Foundation University Professor and Professor of Geography at the University of South Carolina. His specialties include fluvial geomorphology and hydrology, as well as policy for public land and water. His PhD is from the University of Wisconsin, Madison, with a major in physical geography and a minor in water resources management. His research and teaching have focused on river-channel change, human impacts on river processes, morphology, and ecology, along with contaminant transport and storage in river systems. In the arena of public policy, he has emphasized the interaction of science and decision making, and the resolution of conflicts among economic development, historical preservation, and environmental restoration for rivers, particularly river restoration for endangered species. He has authored or edited eight books, more than 125 scientific papers, book chapters, and reports, more than 60 successful grant proposals, and more than 100 public presentations. He has been an officer in the Geological Society of America. President Clinton appointed him to the Presidential Commission on American Heritage Rivers. His National Research Council (NRC) Service includes membership of the Water Science and Technology Board and the Board of Earth Sciences and Resources. He has chaired four NRC committees and is a National Associate of the National Academies. Graf was President of the AAG in 1998–99.

SUSAN HANSON is the Jan and Larry Landry University Professor at Clark University's School of Geography. She is an urban geographer and feminist geographer, with interests in the geography of everyday life in cities and access to opportunity. Her BA is from Middlebury College (1964) and MS (1969) and PhD (1973) are from Northwestern University. Hanson has edited *The Professional Geographer*, the *Annals of the Association of American Geographers*, and *Economic Geography*. She is a member of the National Academy of Sciences and the American Academy of Arts and Sciences. Hanson was President of the AAG in 1990-91.

JOHN FRASER HART is Professor of Geography at the University of Minnesota. He was born in Staunton, VA in 1924. He earned his BA at Emory University (1943) and PhD at Northwestern University (1950). Hart held appointments at the University of Georgia (1949-1955), Indiana University (1955–1967) and the University of Minnesota (1967–present). In 1965-66 he served as Executive Director of the Association of American Geographers. He was Editor of the *Annals of the Association of American*

Geographers (1970–1975) and President of the Association in 1979-80. Among Hart's books are *The South* (1967), *The Look of the Land* (1975), *The Rural Landscape* (1998), and *The Changing Scale of American Agriculture* (2003). Hart was President of the AAG in 1979-1980.

NICHOLAS HELBURN completed his BA (Chicago) and PhD (Wisconsin) in geography. In between, he earned a Masters in Agricultural Economics from Montana State University in Bozeman where he fell in love with the Rocky Mountains. He was a conscientious objector in World War II serving as a parachute firefighter with the US Forest Service. Helburn founded the Department of Geography at Montana State in 1947, which he chaired until 1964. He then moved to the University of Colorado, Boulder to work with the High School Geography Project that he directed until 1969. After a year at Western Michigan University, he joined the faculty at the University of Colorado, Boulder, retiring in 1990. He now lives on a small farm in Boulder County raising fruit, vegetables, and pasture. Helburn was President of the AAG in 1980–81.

TERRY G. JORDAN-BYCHKOV (1938–2003), a sixth-generation Texan, attended Southern Methodist University, the University of Texas at Austin, and the University of Wisconsin at Madison, where he received his doctorate in 1965. After teaching at Arizona State University and serving as department chair at the University of North Texas, he took up the Walter Prescott Webb Chair of History and Ideas at the University of Texas, Austin in 1982. He authored over 25 books and monographs and numerous research papers on pioneers, frontiers, and the diffusion of material folk culture and ways of life in the American West, Siberia, Australia, and Europe. His influential textbook, *The Human Mosaic*, first published in 1976, has remained in press with frequent updates. Jordan served as President of the AAG from 1987-88 and during this time used the *AAG Newsletter* as a bully pulpit to argue for the importance of research and scholarly publications as the primary mission of the AAG. His *Newsletter* articles generated a blizzard of correspondence that always received detailed replies.

ROBERT W. KATES is an independent scholar in Trenton, Maine, and University Professor (Emeritus) at Brown University. Having failed retirement, he is Co-Convenor of the International Initiative for Science and Technology for Sustainability; an Executive Editor of *Environment* magazine, Distinguished Scientist, George Perkins Marsh Institute, Clark University, Faculty Associate, College of the Atlantic; Visiting Scholar of the Center for International Development, Kennedy School of Government, Harvard University; and a Board member of the Acadia Disposal District and Maine Global Climate Change, Inc. He is a recipient of the 1991 National Medal of Science, the MacArthur Prize Fellowship (1981–85), and Laureat d'Honneur, International Geographical Union (2004). Kates is a member of the National Academy of Sciences, the American Academy of Arts and Sciences, and the Academia Europaea. Kates is the author, editor, and co-editor of 22 books and monographs. His most recent books are: *Great Transition: The Promise and Lure of the Times Ahead* (2002); and with the AAG Global Change and Local Places Research Group, *Global Change in Local Places: Estimating, Understanding, and Reducing Greenhouse Gases* (2003). Kates was President of the AAG in 1993–94.

CLYDE F. KOHN (1911–1989) began his career as an elementary and secondary teacher, and in teacher education. His commitments to geographic education were sustained throughout his professional life. Kohn earned his BA at Northern Michigan and his MA and PhD at the University of Michigan. His two long-term faculty appointments were at Northwestern (1948–58) and at the University of Iowa (1958–84) where he chaired the Department of Geography from 1965–78. He served in many leadership roles for the National Council for Geographic Education from which he received the Distinguished Service Award (1973) and Distinguished Mentor Award (1985). In the AAG he chaired and served on an array of education-related committees, helping to initiate the path-breaking High School Geography Project. He wrote or edited more than 20 books including school texts, consulted for educational media, and published numerous research articles. Kohn was President of the AAG in 1967–68.

PEIRCE LEWIS is Professor Emeritus of Geography at the Pennsylvania State University. For four decades he has been a student of American physical and human landscapes, especially the ordinary urban and rural landscapes created by Americans. He earned his PhD at the University of Michigan in 1958. His most notable publication is *New Orleans: The*

Making of an Urban Landscape. His writings have received awards from the Association of American Geographers and the International Geographical Union. Lewis has been awarded fellowships from the Guggenheim Foundation, the Woodrow Wilson International Center for Scholars, and the Smithsonian Institution. He has received numerous honors for his teaching, including the Lindback Foundation Award, Penn State's highest award for distinguished teaching. Lewis has worked extensively to cross traditional academic boundaries, serving as a consultant to the National Rural Studies Committee and as a contributor to the multidisciplinary study of material culture in the National Museum of American History at the Smithsonian Institution. Lewis was President of the AAG in 1983-84.

MELVIN G. MARCUS (1929–1997) early developed an interest in the outdoors, which led him to complete 52 mountaineering first ascents during his youth and to enter Yale University as a geology major in 1947. His interest in geography blossomed during military service in Korea and Japan. Marcus completed his BA (Miami), MS (Colorado), and PhD (Chicago) in geography. He held faculty positions at Rutgers University (1960-64), the University of Michigan (1964–74), and Arizona State University, where he served until his death. He was department head at both Michigan and Arizona State. He also developed close ties to Canterbury University (New Zealand) and to the US Military Academy at West Point. Throughout his career, his research, publication, teaching and service focused on glaciers, mountain climatology, human impacts on the environment, and geographic and environmental education. His wife, Mary Ann, was a long-time partner in the field. Marcus was Vice President of the American Geographical Society (1986–96) and on the Board of the Yosemite National Institutes (1979–97), a field-based environmental education center that serves over 30,000 children annually. He believed his greatest contribution was introducing students to geography and, in particular, bringing female students to the previously male bastion of field research in remote mountain environments. In 1997 he received both the Cullum Geographical Medal of the AGS and the Lifetime Achievement Award of the AAG. Marcus was President of the AAG in 1977–78.

JOHN RUSSELL MATHER (1923–2002) was born in Boston. After serving as a weather forecaster in World War II, he earned his BA (1945) at

Williams College and a BS (1947) and MS (1948) in Meteorology from
MIT, and PhD from Johns Hopkins (1951). His career began at the C.
Warren Thornthwaite Laboratory of Climatology where he remained for
16 years, becoming president on Thornthwaite's death. In 1972, he joined
the Department of Geography at the University of Delaware after several
years of part-time affiliation, building a strong focus on climatology.
Mather is probably best known for his research on water budgets, but he
published extensively on a variety of topics, including his major historical
contribution, *The Genius of C. Warren Thornthwaite, Climatologist-
Geographer* (co-authored with Marie Sanderson). Among the offices he
held were Council Member of the American Geographical Society (1981–
2000) and its Secretary (1982–2000); and State Climatologist for Delaware
(1978–91). Mather received many awards: including the Commander's
Award for Public Service, US Department of the Army (1990); the Francis
Alison Award for distinguished scholarship at the University of Delaware
(1991); the AAG Lifetime Achievement Honors (1998); and the American
Geographical Society's Charles P. Daly Award (1999). Mather was President
of the AAG in 1991–92.

MARVIN W. MIKESELL, Professor of Geography at the University of Chicago,
received his PhD from the University of California, Berkeley where he was
a student of Carl Sauer. His major interests have been in cultural-political
geography, environmental studies, and the history of ideas in geography
and related fields. His AAG service includes National Councilor (1972–74),
Editor of the Monograph Series (1966–72), and Delegate of the Association
to the US National Commission for UNESCO. Mikesell was President of
the AAG in 1975–76.

JANICE JONES MONK is Professor of Geography and Regional Development
and Research Professor in the Southwest Institute for Research on
Women (SIROW) at University of Arizona. She earned her BA at the
University of Sydney and her MA and PhD at the University of Illinois
at Urbana-Champaign. Her publications are in feminist geography,
including the history of women in American geography, and geographic
education, especially in higher education. Monk's awards include Lifetime
Achievement Honors (2000) and Honors (1992) from the Association of
American Geographers, the George J. Miller Distinguished Service Award

(1998) from the National Council for Geographic Education, the Australia-International Medal from the Institute of Australian Geographers (1999), and the Taylor and Francis Award of the Royal Geographical Society (with the Institute of British Geographers (2003). She has strong interests in international collaboration and has been a visiting professor/fellow in Australia, India, The Netherlands, New Zealand, Republic of China, and Spain. She was President of the AAG in 2001–02.

RICHARD MORRILL, a native of Los Angeles in a liberal family, was early on a social and political activist. He became a geographer at Dartmouth (1951-55), moving to the University of Washington in 1955. By 2005 he will have loved and taught geography for 50 years. Morrill was an enthusiastic player in the theoretical/quantitative/modeling revolution led by Garrison and he remains a fervent exponent of geography as a science of spatial behavior. In 1959 he was invited to "spread the word" to Northwestern and next went to Sweden on a National Science Foundation grant. Morrill has since remained at the University of Washington except for brief intervals, for example as Director of the Chicago Regional Hospital Study. His work in economic, social, population, transportation, medical, and political geography may seem highly eclectic, but is linked by interest in spatial behavior and commitment to social justice, often devoted to themes like inequality, race and segregation, access to health care, or political gerrymandering. Morrill has enjoyed many leadership opportunities including Presidency of the Western Regional Science Association. His awards include a Guggenheim Fellowship. He has nevertheless found opportunities for escape—to walk, hike, and climb, and enjoys a great family and a brilliant editor wife. Morrill was President of the AAG in 1981–82.

ALEXANDER B. MURPHY is Professor of Geography at the University of Oregon, where he also holds the James F. and Shirley K. Rippey Chair in Liberal Arts and Sciences. He earned his BA (Archeology) at Yale, JD at Columbia, and PhD in Geography at the University of Chicago. Murphy specializes in political and cultural geography, with a regional emphasis on Europe. He is a vice president of the American Geographical Society and an editor of *Progress in Human Geography and Eurasian Geography and Economics*. Murphy is the author or co-author of more than 60 articles and several books, including *The Regional Dynamics of Language Differentiation in Belgium*; *Cultural Encounters with the Environment*, (with Douglas Johnson); and *Human Geography: Culture, Society, and Space*, (with Harm de Blij). He has received a Fulbright-Hays Research Grant (1985), a National Endowment for the Humanities Fellowship (1991), the Ersted Award for Distinguished Teaching at the University of Oregon (1991), a National Science Foundation Presidential Young Investigator Award in the mid–1990s, and the National Council for Geographic Education Distinguished Teaching Award (2001). He served as President of the AAG in 2003-04.

M. DUANE NELLIS is currently Provost and Professor of Geography at Kansas State University, a position he has held since June 2004. Prior to his appointment at Kansas State University, he served seven years as Dean of the Eberly College of Arts and Sciences at West Virginia University. He is past President of the National Council for Geographic Education, and past President of Gamma Theta Upsilon, the international geography honor society. He has published over 100 articles and 15 books and book chapters on various aspects of rural geography and the applications of geospatial technology to rural land use systems. Nellis has received numerous awards for his work including Honors from the AAG, Outstanding Contributions from the Remote Sensing Specialty Group of the AAG, and the John Fraser Hart Award for Research Excellence. He is a member of the Research and Graduate Studies Subcommittee of the Council of Academic Affairs for the National Association of State Universities and Land Grant Colleges. Nellis was President of the AAG in 2002–03.

JUDY M. OLSON is Professor of Geography at Michigan State University, where she has been since 1983, and serving as Department Chair from 1989-94. She received her bachelor's degree (1966) from the University of Wisconsin—Stevens Point, and her Master's (1968) and PhD (1970) from the University of Wisconsin—Madison. She was on the faculties of the University of Georgia and Boston University before coming to MSU. She has been active in the American Congress on Surveying and Mapping and served as Vice President of the International Cartographic Association (ICA) and Chair of US National Committee for ICA. She edited *The American Cartographer* (now *Cartography and Geographic Information Science*) for six years and has served on various editorial boards, NSF panels, and program/department review committees. Her research and teaching interests have been in cartography, geographic information systems, quantitative methods, and, in recent years, geographic education. She has published in the *Annals of the Association of American Geographers, The Professional Geographer, The American Cartographer, The Cartographic Journal, The International Yearbook of Cartography, Canadian Cartographer* (now *Cartographica*), and *Journal of Geography*. Her favorite classes read widely in the discipline, do creative research, and produce thought-provoking graphics. Olson was President of the AAG in 1995–96.

RISA PALM is currently Professor of Geography and also Executive Vice Chancellor and Provost at Louisiana State University. She has held several other administrative positions including Dean of Arts and Sciences at the Universities of North Carolina (Chapel Hill) and Oregon, and Associate Vice Chancellor for Research and Dean of the Graduate School at the University of Colorado, Boulder. Palm received her PhD from the University of Minnesota. She has held faculty positions at Normandale Community College (Bloomington, Minnesota) as well as at the University of California, Berkeley. Her research has focused on intra-urban migration and the components of the housing market, as well as on societal response to earthquake hazards. She is the author or co-author of thirteen books and monographs, and has received several major grants from the National Science Foundation to support empirical research on the response of the housing market and the risk assessment by households to earthquake hazards in California, Puerto Rico, and Japan. Her most recent publications investigated the use of the internet as an information source in the intra-urban migration process, and also the economic and population correlates of international telephone connectivity. She has received the Honors Award from the Association of American Geographers and also a Lifetime Achievement Award from the Southeast Division of the AAG. She is a member of the Executive Committee of the Council in Academic Affairs for the National Association of State Universities and Land Grant Colleges. Palm was President of the AAG in 1984–85.

JAMES J. PARSONS (1915–1997) was Professor of Geography at the University of California, Berkeley where he also earned his PhD, taught from 1940 until his official retirement in 1986, and maintained strong associations for the rest of his life. During this career he chaired 38 doctoral committees and encouraged many other geographers. A distinguished Latin Americanist, Parsons published extensively for almost 60 years especially in *The Geographical Review*, on themes in historical geography, landscape, nature, and place. *Hispanic Lands and Peoples: Selected Writings of James J. Parsons* (ed. W.M. Denevan, 1989) reflect his enduring contributions. Parsons was President of the Association of Pacific Coast Geographers (1954–55). His awards include a Guggenheim Fellowship, Honors from the Association of American Geographers (1983) and recognition from government and institutions in Colombia. Parsons was President of the AAG in 1974–75.

THOMAS J. WILBANKS is a Corporate Research Fellow at Oak Ridge National Laboratory (ORNL) and leads its Global Change and Developing Country programs. After receiving his PhD in geography in 1969, Wilbanks served on the faculties of Syracuse University and the University of Oklahoma before joining ORNL in 1977. The designation "Corporate Fellow" is roughly equivalent to a chaired professorship in a university, limited to about 25 individuals from a research staff of about 150. Between 1992–1997 he chaired both of geography's committees in the National Academy of Sciences/National Research Council (NAS/NRC), one of which produced the first NAS assessment of geography in a quarter-century. He has received AAG Honors (1986), the Distinguished Geography Educator's Award of the National Geographic Society (1993), the James R. Anderson Medal of Honor in Applied Geography (1995), and was elected a Fellow of the American Association for the Advancement of Science (1985). Wilbanks is a member of the Board on Earth Sciences and Resources of the NRC and of its Committee on the Human Dimensions of Global Change. He is a member of the Science Steering Group for the US Carbon Cycle Program and been active in a number of other scientific committees related to global change. Wilbanks was President of the AAG in 1992–93.

DAVID WARD became the 11th president of the American Council on Education in September 2001 after serving for eight years as Chancellor of the University of Wisconsin-Madison where he received his PhD in 1963. Ward also served as Associate Dean of the Graduate School (1980–87) and as Vice Chancellor for Academic Affairs and Provost (1989–1993). Ward's service to higher education includes the chairmanship of the Board of Trustees of the University Corporation for Advanced Internet

Development, a non-profit group that spearheaded the development of Internet 2 and membership of the Kellogg Commission on the Future of State and Land-Grant Universities. He also held the Andrew Hill Clark Professorship of Geography at the University of Wisconsin-Madison and chaired the Department of Geography (1974–1977). As an urban geographer, he pioneered research in English and American cities during their rapid industrialization in the 19th and early 20th centuries. He has held visiting appointments at University College London; the Australian National University; Hebrew University, and at his undergraduate alma mater, the University of Leeds. Ward was President of the AAG in 1989-90.

WILBUR ZELINSKY, a native of Chicago, attended five colleges while discovering that he was meant to be a geographer before earning a doctorate at his sixth institution, the University of California, Berkeley. Three of his most influential mentors have been the redoubtable Carl Sauer, Glenn Trewartha, and John K. Wright. After an abortive fling at cartography, and appointments at the University of Georgia and Southern Illinois University, he has been firmly and happily anchored at the Pennsylvania State University since 1963. Major research interests have been in population and historical geography, but most especially along the unexplored borders of the social and cultural geography of North America. He has participated in the 50th, 75th, and 100th anniversary meetings of the Association of American Geographers and has every intention of making it to the 125th. Zelinsky was President of the AAG in 1972–73.

***